T0239805

Engineering Mathematics with Applications to Fire Engineering

Engineering Mathematics with Applications to Fire Engineering

Khalid Khan
Tony Lee Graham

CRC Press
Taylor & Francis Group
Boca Raton London New York

CRC Press is an imprint of the
Taylor & Francis Group, an **informa** business

CRC Press
Taylor & Francis Group
6000 Broken Sound Parkway NW, Suite 300
Boca Raton, FL 33487-2742

First issued in paperback 2020

© 2018 by Taylor & Francis Group, LLC
CRC Press is an imprint of Taylor & Francis Group, an Informa business

No claim to original U.S. Government works

ISBN-13: 978-0-367-57147-4 (pbk)
ISBN-13: 978-1-138-09884-8 (hbk)

This book contains information obtained from authentic and highly regarded sources. Reasonable efforts have been made to publish reliable data and information, but the author and publisher cannot assume responsibility for the validity of all materials or the consequences of their use. The authors and publishers have attempted to trace the copyright holders of all material reproduced in this publication and apologize to copyright holders if permission to publish in this form has not been obtained. If any copyright material has not been acknowledged please write and let us know so we may rectify in any future reprint.

Except as permitted under U.S. Copyright Law, no part of this book may be reprinted, reproduced, transmitted, or utilized in any form by any electronic, mechanical, or other means, now known or hereafter invented, including photocopying, microfilming, and recording, or in any information storage or retrieval system, without written permission from the publishers.

For permission to photocopy or use material electronically from this work, please access www.copyright.com (http://www.copyright.com/) or contact the Copyright Clearance Center, Inc. (CCC), 222 Rosewood Drive, Danvers, MA 01923, 978-750-8400. CCC is a not-for-profit organization that provides licenses and registration for a variety of users. For organizations that have been granted a photocopy license by the CCC, a separate system of payment has been arranged.

Trademark Notice: Product or corporate names may be trademarks or registered trademarks, and are used only for identification and explanation without intent to infringe.

Library of Congress Cataloging-in-Publication Data

Names: Khan, Khalid Mahmood, 1963- author. | Graham, Tony Lee, author.
Title: Engineering mathematics with applications to fire engineering / Khalid Khan and Tony Lee Graham.
Description: Boca Raton : Taylor & Francis, 2018. | Includes bibliographical references and index.
Identifiers: LCCN 2018001261| ISBN 9781138098848 (hardback : acid-free paper) | ISBN 9781315104270 (ebook)
Subjects: LCSH: Engineering mathematics. | Engineering mathematics--Problems, exercises, etc. | Fire protection engineering--Mathematics.
Classification: LCC TA332.5 .K43 2018 | DDC 620.001/51--dc23
LC record available at https://lccn.loc.gov/2018001261

Visit the Taylor & Francis Web site at
http://www.taylorandfrancis.com

and the CRC Press Web site at
http://www.crcpress.com

Contents

Preface

Engineering is the application of scientific knowledge to solving problems in the real world. There are many branches and fields within the different engineering disciplines. All disciplines, whether they be mechanical, civil, chemical, electrical, or fire, have common skills that rely heavily on rational thinking and logical decision making. A very important component required in all engineering fields is a good understanding of applied mathematical concepts.

This book was written for undergraduate degree–level students studying a first course in engineering. Its primary aim is students studying fire engineering and related courses, although there are many other real-world applications given that are relevant to other areas of engineering. The motivation for writing such a book was that although there are many engineering mathematics textbooks available for fire engineering students, there was not a book that showed how the mathematical concepts studied were applied to their particular discipline. This book not only gives the foundation mathematics needed for fire engineering degree courses but also gives real-world fire applications showing how these concepts are used in practice.

This book is based on a combination of lectures that have been delivered over a twenty-year period at international educational institutes and at the University of Central Lancashire in the United Kingdom across the various levels of undergraduate study. The Division of Fire and Safety Engineering at the University of Central Lancashire has developed a leading position in the United Kingdom and overseas with respect to the delivery of undergraduate and postgraduate courses in fire safety engineering, management, and leadership studies. The step-by-step methodology used together with plenty of practical applications in the real world make this book an essential aid in the understanding of mathematical concepts for most engineering disciplines. Other engineering mathematics textbooks generally assume a certain level of mathematical understanding and therefore sometimes certain key steps are omitted in the discourse; this is not the case with the current book.

Each chapter of the book has a similar format starting with the development of the basic mathematical theory, which is then followed by real-world applications especially to fire engineering, and then concluding with a problems section.

We consider the presentation in this text as unique in that there are real-world fire engineering applications given alongside the mathematical content that underpins those applications.

The overall structure of the book is that it begins with a review of the key elementary mathematical concepts and focuses on important concepts such as transposition of formulae, as these form an essential part of many engineering solutions. An introduction to probability theory with discrete and continuous random variables are important concepts used in fault tree analysis in risk assessment and reliability theory, respectively. Determinants and matrices lead to solving a system of linear equations using different techniques such as Gaussian elimination and the matrix inversion method. Vectors and normal vectors to surfaces are considered and form the basis of concepts in surface integrals. Use of complex numbers and their applications in electrical circuit theory and the role they play in the solutions to differential equations are covered. There is an introduction to one-variable calculus with the fundamentals of differentiation, integration techniques, and importance of integration as a summing process for a multitude of engineering applications. Methods of solving ordinary linear differential equations are introduced with emphasis on the Laplace transform method as a valuable tool in finding solutions to problems. Higher-dimensional multivariable calculus dealing with partial derivatives, double and triple integrals, and general change of coordinate systems are covered. Finally, in the last chapter on vector calculus, vector fields representing physical phenomenon are considered along with concepts of divergence and curl, and applications of Green's, Stokes', and divergence theorems are given.

We would like to thank the staff at CRC/Taylor & Francis who have contributed to the production of this book and to Dr. Alan Burns for his valuable comments on the draft version of the book. Finally, we wish to express our gratitude to the University of Central Lancashire for providing a conductive academic environment that allowed us to complete this project.

Authors

Khalid Khan, BSc (Hons), MSc, PhD, received his BSc (Hons) in mathematical physics, and MSc and PhD in control systems all from the University of Manchester Institute of Science and Technology in the United Kingdom. Dr Khan then spent two years as a consultant engineer working on safety, reliability and risk assessment problems in the energy industry. Subsequently, he moved abroad and after spending some period of time as an assistant professor of mathematics at Etisalat University in the United Arab Emirates he returned to the United Kingdom and took up a position at the University of Central Lancashire.

Dr Khan is currently a senior lecturer in engineering mathematics in the School of Engineering at the University of Central Lancashire. He teaches on a range of mathematics modules within the fire degree programs and contributes to other mathematics teaching within the school and college. He is currently the course leader for the Foundation Degree in Fire Safety Engineering. Dr Khan as part of the fire team has been involved in the development of a range of courses in fire safety engineering and management that are currently running at one of University of Central Lancashire's international partnerships in Qatar. Dr Khan is also a senior fellow of the Higher Education Academy (HEA), which is a British professional institution promoting excellence in higher education.

Dr Khan's research interests are in the area of mathematical modeling of system behavior in a range of applications. His current work focuses on fire suppression using sprinkler systems and on mathematical models of collective motion of self-propelled particles in homogeneous and heterogeneous mediums. Dr Khan currently has over thirty publications and is also a member of two journal review panel boards.

Tony Graham, BSc (Hons), PhD, is a senior lecturer and course leader at the University of Central Lancashire, United Kingdom. He is best known for teaching fire safety engineering to thousands of students over twenty-four years in different countries and for papers on compartment fire dynamics and the phenomenon of flashover fire. His teaching also includes risk engineering and engineering analysis. He has taught courses at International College of Engineering and Management in Sultanate of Oman and also at City University of Hong Kong.

He is a senior fellow of Higher Education Academy and Secretary of the Combustion, Explosion and Fire Engineering research group in the School of Engineering and has served on both the Academic Board, and Academic Standards and Quality Assurance Committee. His research was funded by Nuffield Foundation and is ongoing. Dr Graham earned his BSc (Hons) in theoretical physics at Lancaster University, and subsequently was awarded his PhD in fire engineering in 1998 from the University of Central Lancashire. For the future, Dr Graham is looking at two books on fire engineering science and risk engineering. Dr Graham, along with his beloved wife Anna and daughter Natalia now prosper in Preston, Lancashire.

1 Review of Basic Concepts

1.1 Degrees of Accuracy

1.1.1 Rounding Numbers (Common Method)

In the real world, when dealing with numbers a degree of accuracy is needed. For example, if a piece of wood of a certain length was required, asking for a length of 123.732461 centimeters would not be sensible as such accurate measurements are not possible. What is more usual is some form of rounding. The method of rounding is commonly used in mathematical applications in science and engineering. It is the one generally taught in mathematics classes in high school. The method is also known as *round-half-up*. It works as follows:

- Decide which is the last digit to keep.

- Increase it by 1 if the next digit is 5 or more (this is called rounding up).

- Leave it the same if the next digit is 4 or less (this is called rounding down).

Example 1.1

3.044 rounded to hundredths is 3.04 (because the next digit, 4, is less than 5).

3.045 rounded to hundredths is 3.05 (because the next digit, 5, is 5 or more).

3.0447 rounded to hundredths is 3.04 (because the next digit, 4, is less than 5).

For negative numbers, one rounds the *absolute value* and reapplies the sign afterward.

Example 1.2

−2.1349 rounded to hundredths is −2.13.
−2.1350 rounded to hundredths is −2.14.

1.1.2 Round-to-Even Method

The round-to-even method method, also known as *unbiased rounding* or *Gaussian rounding*, exactly replicates the common method of rounding *except when the digit(s) following the rounding digit starts with a 5 and has no nonzero digits after it*. The new algorithm becomes

- Decide which is the last digit to keep.

- Increase it by 1 if the next digit is 6 or more, or a 5 followed by one or more nonzero digits.

- Leave it the same if the next digit is 4 or less.

- Otherwise, if all that follows the last digit is a 5 and possibly trailing zeros, then increase the rounded digit if it is currently odd; else, if it is already even, leave it alone.

All rounding schemes have two possible outcomes: increasing the rounding digit by one or leaving it alone. With traditional rounding, if the number has a value less than the halfway mark between the possible outcomes, it is rounded down; if the number has a value exactly halfway or greater than halfway between the possible outcomes, it is rounded up. The round-to-even method is the same except that numbers exactly halfway between the possible outcomes are sometimes rounded up, sometimes down.

Despite the custom of rounding the number 4.5 up to 5, 4.5 is no nearer to 5 than it is to 4 (it is 0.5 away from both). When dealing with large sets of scientific or statistical data, where trends are important, traditional rounding on average biases the data upward slightly. Over a large set of data, or when many subsequent rounding operations are performed as in *digital signal processing*, the round-to-even rule tends to reduce the total rounding error, with (on average) an equal portion of numbers rounding up as rounding down. This generally reduces upward skewing of the result.

Examples 1.3

3.016 rounded to hundredths is 3.02 (because the next digit, 6, is 6 or more).
3.013 rounded to hundredths is 3.01 (because the next digit, 3, is 4 or less).
3.015 rounded to hundredths is 3.02 (because the next digit is 5, and the hundredths digit, 1, is odd).
3.045 rounded to hundredths is 3.04 (because the next digit is 5, and the hundredths digit, 4, is even).
3.04501 rounded to hundredths is 3.05 (because the next digit is 5, but it is followed by nonzero digits).

Table 1.1 Showing Accuracy of the "Round-to-Even" Method

Original Number	"Old" Rounding Method	"Round-to-Even" Method
3.55	3.6	3.6
3.65	3.7	3.6
3.75	3.8	3.8
3.85	3.9	3.8
3.95	4.0	4.0
Mean = 3.75	Mean = 3.8	Mean = 3.76

Example 1.4

Table 1.1 shows how the round-to-even system compares with the old system of rounding. There are five original data points that are rounded to the tenths and then their average is calculated. It can be seen that the round-to-even method is much more accurate than the old method of rounding.

1.1.3 Decimal Places

When rounding a number, one is usually told how to round it. It is simplest when one is told how many places to round to, but one should also know how to round to a named place, such as to the nearest thousand or to the ten-thousandths place. Also, it may be required to know how to round to a certain number of significant digits; this is dealt with later.

Using the first few digits of the decimal expansion of $pi = \pi = 3.14159265...$ in the following examples.

Example 1.5

Round pi to five decimal places. First, count out the five decimal places, then look at the sixth place:

$$3.14159|265...$$

A little line separating the fifth place from the sixth place has been drawn. This can be a good way of keeping your place, especially if dealing with lots of digits.

The fifth place has a 9 in it. Looking at the sixth place, it has a 2 in it. Since 2 is less than 5, the 9 will not be rounded up; that is, just leave the 9 as it is. In addition, delete the digits after the 9. Then pi, rounded to five decimal places, is given as 3.14159.

Example 1.6

Round pi to four decimal places. First, go back to the original number: 3.14159265. Count off four places, and look at the number in the fifth place:

$$3.1415|9265...$$

The number in the fifth place is a 9, which is greater than 5, so round up in the fourth place, truncating the expansion at four decimal places.

That is, the 5 becomes a 6, the 9265... part disappears, and *pi*, rounded to four decimal places, is given as 3.1416.

This rounding works the same way when rounding to a certain named place, such as the hundredths place. The only difference being a bit more careful in counting off the places needed. Just remember that the decimal places count off to the right in the same order as the counting numbers count off to the left. That is, for regular numbers, the place values are

...(ten-thousands) (thousands) (hundreds) (tens) (ones)

For decimal places, a "oneths" is not there, but all the other fractions are

(decimal point) (tenths) (hundredths) (thousandths) (ten-thousandths)...

Example 1.7

Round *pi* to the nearest thousandth. The "nearest thousandth" means that one needs to count off three decimal places (tenths, hundredths, thousandths), and then round:

$$3.141|59265...$$

Then *pi*, rounded to the nearest thousandth, is 3.142.

Example 1.8

Round 18.796 to the hundredths place. The hundredths place is two decimal places, so count off two decimal places, and round according to the third decimal place:

$$18.79|6$$

Since the third decimal place contains a 6, which is greater than 5, one has to round up. But rounding up a 9 gives a 10. In this case, round the 79 up to an 80 as 18.80.

One might be tempted to write this as 18.8, but, since rounded to the hundredths place (to two decimal places), one should write both decimal places. Otherwise, it looks like rounding to one decimal place, or to the tenths place, and the answer could be counted off as being incorrect.

1.1.4   Significant Places

Rounding can also be carried out to an appropriate number of significant digits. What are significant digits? Well, they are sort of the "interesting" or "important" digits. For example:

3.14159 has six significant digits (all the numbers give you useful information).

1000 has one significant digit (only the 1 is interesting; you do not know anything for sure about the hundreds, tens, or units places; the zeroes may just be placeholders; they may have rounded something off to get this value).

1000.0 has five significant digits (the ".0" tells us something interesting about the presumed accuracy of the measurement being made: that the measurement is accurate to the tenths place, but that there happens to be zero tenths).

0.00035 has two significant digits (only the 3 and 5 tell us something; the other zeroes are placeholders, only providing information about relative size).

0.000350 has three significant digits (that last zero tells us that the measurement was made accurate to that last digit, which just happened to have a value of zero).

1006 has four significant digits (the 1 and 6 are interesting, and the zeros have to be counted, because they are between the two interesting numbers).

560 has two significant digits (the last zero is just a placeholder).

560. (notice the point after the zero) has three significant digits (the decimal point tells us that the measurement was made to the nearest unit, so the zero is not just a placeholder).

560.0 has four significant digits (the zero in the tenths place means that the measurement was made accurate to the tenths place, and that there just happen to be zero tenths; the 5 and 6 give useful information, and the other zero is between significant digits, and must therefore also be counted).

Here are the basic rules for significant digits:

1. All nonzero digits are significant.

2. All zeros between significant digits are significant.

3. All zeros that are both to the right of the decimal point and to the right of all nonzero significant digits are themselves significant.

Following are some rounding examples; each number is rounded to four, three, and two significant digits.

Example 1.9

Round 742,396 to four, three, and two significant digits:

742,400 (four significant digits)
742,000 (three significant digits)
740,000 (two significant digits)

Example 1.10

Round 0.07284 to four, three, and two significant digits:

  0.07284 (four significant digits)
  0.0728 (three significant digits)
  0.073 (two significant digits)

Example 1.11

Round 231.45 to four, three, and two significant digits:

  231.4 (four significant digits)
  231 (three significant digits)
  230 (two significant digits)

1.2   Scientific Notation (Standard Form)

Scientific notation, also sometimes known as *standard form* or as *exponential notation*, is a way of writing numbers that accommodates values too large or small to be conveniently written in standard decimal notation. Scientific notation has a number of useful properties and is often favored by engineers, scientists, and mathematicians, who work with such numbers.

Look at the following numbers:

  300,000,000 m/sec, the speed of light

  0.000 000 000 753 kg, mass of a dust particle

Scientists have developed a shorter method to express very large and very small numbers. This method is called scientific notation. Scientific notation is based on powers of the base number 10.

Example 1.12

Our galaxy to which the sun belongs is called the Milky Way. It contains at least 100,000,000,000 stars. Now let's look at this number: 100,000,000,000. It can be written as $1.0 \times 100,000,000,000$. It is the large number 100,000,000,000 that causes the problem. But that is just a multiple of 10. In fact, it is 10 times itself 11 times:

$$10 \times 10 \times 10 \times 10 \times 10 \times 10 \times 10 \times 10 \times 10 \times 10 \times 10 = 100,000,000,000$$

A more convenient way of writing 100,000,000,000 is 10^{11}. The small number to the right of the 10 is called the exponent, or the power of ten. It represents the number of zeros that follow the 1.

So, one would write 100,000,000,000 in scientific notation as 1.0×10^{11}. This number is read as follows: one point zero times ten to the eleventh power.

So generally, any number can be written in scientific form as

$$A \times 10^N \hspace{3cm} (1.1)$$

where $1 \leq A < 10$ and N is any positive or negative integer.

1.2.1   How Does Scientific Notation Work?

As stated earlier, the exponent refers to the number of zeros that follow the 1. So

$$10^1 = 10$$
$$10^2 = 100$$
$$10^3 = 1,000$$

and so on. Similarly, $10^0 = 1$, since the zero exponent means that no zeros follow the 1.

Negative exponents indicate negative powers of 10, which are expressed as fractions with 1 in the numerator (on top) and the power of 10 in the denominator (on the bottom). So

$$10^{-1} = 1/10$$
$$10^{-2} = 1/100$$
$$10^{-3} = 1/1,000$$

and so on. This allows one to express other small numbers this way. For example,

$$2.5 \times 10^{-3} = 2.5 \times 1/1,000 = 0.0025$$

Every number can be expressed in scientific notation. In Example 1.12, 100,000,000,000 should be written as 1.0×10^{11}.

This illustrates another way to think about scientific notation: The exponent will tell you how the decimal point moves; a *positive exponent* moves the decimal point to the *right*, and a *negative* one moves it to the *left*. So, for example,

$$4.0 \times 10^2 = 400 \, (2 \text{ places to the right of } 4)$$

$$4.0 \times 10^{-2} = 0.04 \, (2 \text{ places to the left of } 4)$$

Note that scientific notation is also sometimes expressed as E (for exponent), as in 4 E 2 (meaning 4.0×10 raised to 2). Similarly, 4 E –2 means 4 times 10 raised to –2, or $= 4 \times 10^{-2} = 0.04$. This method of expression makes it easier to type in scientific notation.

1.2.2   Addition and Subtraction

The key to adding or subtracting numbers in scientific notation is to make sure the *exponents are the same*.

Example 1.13

$$(2.0 \times 10^2) + (3.0 \times 10^3)$$

can be rewritten as

$$(0.2 \times 10^3) + (3.0 \times 10^3)$$

Just add 0.2 + 3 and keep the 10^3 intact. Your answer is 3.2 × 10^3, or 3,200. This can be checked by converting the numbers first to the more familiar form. So,

$$2 \times 10^2 + 3.0 \times 10^3 = 200 + 3,000 = 3,200 = 3.2 \times 10^3$$

Example 1.14

$$(2.0 \times 10^7) - (6.3 \times 10^5)$$

The problem needs to be rewritten so that the exponents are the same. So this can be written as

$$(200 \times 10^5) - (6.3 \times 10^5) = 193.7 \times 10^5$$

which in scientific notation would be written as 1.937 × 10^7.

1.2.3   Multiplication

When multiplying numbers expressed in scientific notation, the exponents can simply be added together. This is because the exponent represents the number of zeros following the one. So,

$$10^1 \times 10^2 = 10 \times 100 = 1,000 = 10^3$$

Checking that it is seen $10^1 \times 10^2 = 10^{1+2} = 10^3$.
   Similarly,

$$10^1 \times 10^{-3} = 10^{1-3} = 10^{-2} = .01$$

Again, when checking, it is seen that 10 × 1/1000 = 1/100 = .01.

Example 1.15

Multiply the following numbers: $(4.0 \times 10^5) \times (3.0 \times 10^{-1})$. The 4 and the 3 are multiplied, giving 12, but the exponents 5 and –1 are added, so the answer is 12 × 10^4, or 1.2 × 10^5.
Checking: $(4 \times 10^5) \times (3 \times 10^{-1}) = 400,000 \times 0.3 = 120,000 = 1.2 \times 10^5$.

1.2.4 Division
Example 1.16

$$\frac{(6.0 \times 10^8)}{(3.0 \times 10^5)}$$

To solve this problem, first divide the 6 by the 3, to get 2. The exponent in the denominator is then moved to the numerator, reversing its sign (this will be explained further when dealing with indices). So, move the 10^5 to the numerator with a negative exponent, which then looks like this: $2 \times 10^8 \times 10^{-5}$. All that is left now is to solve this as a multiplication problem, remembering that all that needs to be done for the $10^8 \times 10^{-5}$ part is to add the exponents. So, the answer is 2.0×10^3 or 2,000.

Note: Usually the risk of pedestrian dying in a transport accident is quoted as 1 in 47,773. This then gives the risk of dying as 2.1×10^{-5} (correct to two significant digits). The risk of dying from a fire is quoted as six in a million and can be written as 6×10^{-6}.

1.3 Basic Algebra

1.3.1 Algebraic Notation

Algebraic notation describes how algebra is written. It follows certain rules and conventions, and has its own terminology.

For example, the expression in Figure 1.1 has the following components to it: 1: exponent (power); 2: coefficient; 3: terms; 4: operators; 5: constant, and with x, y: variables.

A *coefficient* is a numerical value or letter representing a numerical constant that multiplies a variable (the operator is omitted).

A group of coefficients, variables, constants and exponents that may be separated from the other terms by the plus and minus operators. Letters represent variables and constants.

By convention, letters at the beginning of the alphabet (e.g., *a*, *b*, and *c*) are typically used to represent *constants*, and those toward the end of the alphabet (e.g., *w*, *x*, and *y*) are used to represent *variables*.

Algebraic operations work in the same way as *arithmetic operations*, such as *addition*, *subtraction*, *multiplication*, *division*, and *exponentiation*, and are applied to algebraic variables and terms.

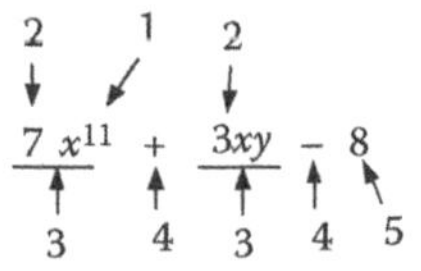

Figure 1.1 Composition of an algebraic expression.

Multiplication symbols are usually omitted, and implied when there is no space between two variables or terms, or when a coefficient is used. For example, $5 \times x^2$ is written as $5x^2$.

1.3.2 Evaluating Expressions

Algebraic expressions may be evaluated and simplified, based on the basic properties of arithmetic operations (addition, subtraction, multiplication, division, and exponentiation).

Added terms are simplified using coefficients. For example, $x + x + x + x$ can be simplified as $4x$ (where 4 is a numerical coefficient).

Multiplied terms are simplified using exponents. For example, $x \times x \times x \times x$ is represented as x^4.

Like terms are added together, for example, $5x + 4p + 1 - 2x + 6p + 4$ is written as $3x + 10p + 5$.

Brackets can be multiplied out, using the *distributive* property. For example, $4(x + 3)$ can be written as $4 \times x + 4 \times 3$, which can be written as $4x + 12$.

This idea can be extended to multiply out two brackets as $(x + 4)(x + 3)$. Here in the first bracket the x term multiplies the $(x + 3)$ and then $+4$ term multiplies the $(x + 3)$ as follows:

$$(x+4)(x+3) = x(x+3) + 4(x+3) = x^2 + 3x + 4x + 12 = x^2 + 7x + 12.$$

Expressions can be factored. For example, $6x + 24x^3$, by taking a factor of $6x$ from both terms can be written as $6x(1 + 4x^2)$.

1.4   Linear Equations

1.4.1   Solving Linear Equations

In engineering, the physical modeling of system behavior is done using the language of mathematics. Different types of equations are derived that model the systems and the simplest type of equations are called linear. An equation of the form $ax + b = 0$, where a and b are constants, is said to be a linear equation with variable x.

Note: For an equation to be linear, the power the variable, in this case x, has to be raised to is one.

The method of solving linear equations is to collect all the terms involving x on one side of the equation and everything else on the other side. The idea is to isolate the variable x to be on its own. The way this is achieved is through using certain operations like addition, subtraction, multiplication, division, and others (i.e., squaring and square rooting) to manipulate the equation so as to keep both sides of the equation the same. This is illustrated in the following examples.

Example 1.17

Solve the following equation to find x: $x + 3 = 7$.
Here it is easy to see what the answer should be for x, $x = 4$. But how can this be arrived at systematically since for more complicated equations

the answer will not be obvious. The approach is to consider the equation as

Left Hand Side (LHS) = Right Hand Side (RHS)

So, to keep this balanced, any operation carried out on the LHS must also be carried out on the RHS. Starting with the equation given earlier:

$x + 3 = 7$ (−3 from both sides of the equation)

$x + 3 - 3 = 7 - 3$ (tidying up both sides)

$x = 4$ (solved for x)

Example 1.18

Solve the following equation to find x: $x - 5 = 4$.
Starting with the equation:

$x - 5 = 4$ (+5 from both sides of the equation)

$x - 5 + 5 = 4 + 5$ (tidying up both sides)

$x = 9$ (solved for x)

Example 1.19

Solve the following equation to find x: $3x = 12$.
Starting with the equation:

$3x = 12$ (divide by 3 on both sides of the equation)

$\dfrac{3x}{3} = \dfrac{12}{3}$ (tidying up both sides)

$x = 4$ (solved for x)

Example 1.20

Solve the following equation to find x: $\dfrac{x}{5} = 6$
Starting with the equation,

$\dfrac{x}{5} = 6$ (multiply by 5 on both sides of the equation)

$\dfrac{x}{5} \times 5 = 6 \times 5$ (tidying up both sides)

$x = 30$ (solved for x)

These basic rules can be applied to more complex equations as follows.

Example 1.21

Solve the following equation to find x: $5x - 3 = 2x + 15$
Starting with the equation:

$5x - 3 = 2x + 15$ (subtract $2x$ from both sides)

$5x - 3 - 2x = 2x + 15 - 2x$ (tidy up both sides)

$3x - 3 = 15$ ($+3$ to both sides)

$3x - 3 + 3 = 15 + 3$ (tidy up both sides)

$3x = 18$ (divide by 3 both sides)

$\dfrac{3x}{3} = \dfrac{18}{3}$ (tidy up both sides)

$x = 6$ (solved for x)

Example 1.22

Solve the following equation to find x: $5(x + 3) + 4(2x - 3) = 2(2x + 15)$
Starting with the equation:

$5(x + 3) + 4(2x - 3) = 2(2x + 15)$ (expand out the brackets)

$5x + 15 + 8x - 12 = 4x + 30$ (tidy up both sides)

$13x + 3 = 4x + 30$ (subtract $4x$ from both sides)

$13x + 3 - 4x = 4x + 30 - 4x$ (tidy up both sides)

$9x + 3 = 30$ (-3 from both sides)

$9x + 3 - 3 = 30 - 3$ (tidy up both sides)

$9x = 27$ (divide by 9 both sides)

$\dfrac{9x}{9} = \dfrac{27}{9}$ (tidy up both sides)

$x = 3$ (solved for x)

Note: Sometimes the linear equations are disguised because they involve fractions. A good strategy is to multiply every term by the lowest common multiple (LCM) of the denominators, as seen in the next example.

Example 1.23

Solve the following equation to find x: $\dfrac{x - 5}{4} - \dfrac{4 - x}{3} = 5$

Starting with the equation:

$$\frac{x-5}{4} - \frac{4-x}{3} = 5$$ (multiply by 12 both sides as this the LCM of 4 and 3)

$$\frac{12(x-5)}{4} - \frac{12(4-x)}{3} = 5 \times 12$$ (tidy up both sides)

$3(x-5) - 4(4-x) = 60$ (expand out brackets)

$3x - 15 - 16 + 4x = 60$ (tidy up both sides)

$7x - 31 = 60$ ($+31$ on both sides)

$7x - 31 + 31 = 60 + 31$ (tidy up both sides)

$7x = 91$ (divide by 7 both sides)

$$\frac{7x}{7} = \frac{91}{7}$$ (tidy up both sides)

$x = 13$ (solved for x)

Note. These basic ideas of solving linear equations are very important in transposing equations.

1.4.2   Transposing Formulae

In science and engineering, formulae are used to relate physical quantities to each other. It is found in electrical circuit theory that the power P is related to the current I and resistance R by the following equation:

$$P = I^2 R$$

Here, P is called the *subject* of the formula.

One may have several sets of corresponding values of I and P and want to find the corresponding values of R. Much time and effort will be saved if the formula was expressed with R as the subject because one then only needs to substitute the given values of I and P in the rearranged formula.

The process of rearranging a formula so that one of the other symbols becomes the subject is called *transposing* the formula. The rules used in the transposition of formulae are essentially the same as those used in solving equations, as seen in the previous section.

Example 1.24: Symbols Connected by a Plus Sign

Transpose $T = t + 10$ to make t the subject.
Rewrite this as $t + 10 = T$.
Now, subtracting 10 from both sides gives $t = T - 10$.

Example 1.25: Symbols Connected by a Minus Sign

Transpose $P - 5V = F$ to make P the subject.
Adding 5V from both sides gives $P = F + 5V$.

Example 1.26: Symbols Connected as Products

Transpose $W = IV$ to make V the subject.
Divide both sides by I:

$$\frac{W}{I} = \frac{IV}{I}$$

or

$$\frac{W}{I} = V \quad \Rightarrow \quad V = \frac{W}{I}$$

Example 1.27: Symbols Connected as a Quotient

Transpose the formula $R = \dfrac{V}{I}$ to make V the subject.
Multiply both sides by I:

$$R \times I = \frac{V}{I} \times I$$

or

$$RI = V \quad \Rightarrow \quad V = RI$$

Example 1.28: Symbols Connected by Multiple Operations

Transpose $V = E + IR$ to make I the subject.
Subtract E from both sides:

$$V - E = IR$$

Divide both sides by R:

$$\frac{V - E}{R} = I \quad \Rightarrow \quad I = \frac{V - E}{R}$$

Example 1.29: Formulae Containing Brackets

Transpose $R = P(1 + 5t)$ to make t the subject.
Remove the bracket on the RHS:

$$R = P + 5Pt$$

Subtract P from both sides:

$$R - P = 5Pt$$

Dividing both sides by 5P:

$$\frac{R-P}{5P} = \frac{5Pt}{5P} \quad \Rightarrow \quad t = \frac{R-P}{5P}$$

Example 1.30: Formulae Containing Roots and Powers

Transpose $J = I^2Rt$ to make I the subject.

Divide both sides by Rt:

$$I^2 = \frac{J}{Rt}$$

Take the square root of both sides:

$$I = \sqrt{\frac{I}{Rt}}$$

Example 1.31

Transpose $x = \sqrt{z-y}$ to make z the subject.

Square both sides:

$$x^2 = z - y$$

Add y to both sides:

$$x^2 + y = z \quad \Rightarrow \quad z = x^2 + y$$

Example 1.32: More Difficult Problem

If $y = \frac{x+2}{x-1}$, transpose to make x the subject.

Multiple by $x - 1$ both sides:

$$(x-1)y = x+2$$

Remove the brackets:

$$xy - y = x + 2$$

Collecting like terms:

$$xy - x = y + 2$$

Take x as a common factor:

$$x(y-1) = y+2$$

Divide by $(y - 1)$:

$$x = \frac{y+2}{y-1}$$

Example 1.33

Alpert's equation for the ceiling jet velocity when the jet is far away from the fire is given as

$$U = 0.195 \left(\frac{\dot{Q}^{1/3} H^{1/2}}{r^{5/6}} \right)$$

where velocity, U, is in meters per second (m/s); total energy release rate, $\dot{Q}$, is in kilowatts (kW); and the ceiling height and radial position (r and H) are in meters (m).

Transpose this equation to make $\dot{Q}$ the subject. First, rewrite to make $\dot{Q}$ on the LHS:

$$0.195 \left(\frac{\dot{Q}^{1/3} H^{1/2}}{r^{5/6}} \right) = U$$

Divide by 0.195 both sides:

$$\left(\frac{\dot{Q}^{1/3} H^{1/2}}{r^{5/6}} \right) = \frac{U}{0.195}$$

Multiply both sides by $r^{\frac{5}{6}}$:

$$\dot{Q}^{1/3} H^{1/2} = \left(\frac{U}{0.195} \right) r^{5/6}$$

Divide both sides by $H^{\frac{1}{2}}$:

$$\dot{Q}^{1/3} = \left(\frac{U}{0.195} \right) \frac{r^{5/6}}{H^{1/2}}$$

Raise both sides to the power 3

$$\dot{Q} = \left[\left(\frac{U}{0.195}\right)\frac{r^{5/6}}{H^{1/2}}\right]^3 = \left(\frac{U}{0.195}\right)^3 \frac{r^{5/2}}{H^{3/2}}$$

Substituting Values into a Formula

In most practical situations, one knows the value of certain variables and constants and is required to calculate the value of another variable. In this case, it may be needed to substitute values into the equation (remembering to use correct units at all times) and calculate the unknown variable.

Example 1.34

If $F = \frac{9}{5}C + 32$, then find F if C = 20.

Solution: Substituting in C = 20 gives

$$F = \frac{9}{5}(20) + 32 \quad \Rightarrow \quad F = 9 \times 4 + 32$$
$$F = 68$$

Now, if one needed to find the value of C given F, then it would be easier to rearrange the equation first to make C the subject and then substitute in the given value of F (see next example).

Example 1.35

Given that $F = \frac{9}{5}C + 32$, find the value of C if F = 86.

Solution: Rearranging

$$\frac{9}{5}C = F - 32$$
$$C = \frac{5}{9}(F - 32)$$

So putting in F = 86 yields

$$C = \frac{5}{9}(86 - 32) \quad \Rightarrow \quad C = 30$$

1.5   Linear Simultaneous Equations

The concern here is primarily with two equations in two variables such as x and y, and this idea can be used to find the intersection of straight lines. There are different methods of solving linear simultaneous equations: elimination method, substitution method, and graphical method. The graphical method is generally not as accurate as the others and is not considered here. The substitution method can sometimes involve awkward fractions and so the elimination method is generally the most preferred method.

1.5.1   Elimination Method

The method of elimination makes the coefficients of one of the variables equal, and then the two equations are either added or subtracted in order to eliminate that variable.

Note: In the equation $3x - 5y = 7$, the coefficients of x and y are 3 and -5, respectively.

Example 1.36

Solve the simultaneous equations

$$x + y = 6 \tag{1}$$

$$x - y = 4 \tag{2}$$

Solution: This is the simplest case, and since the coefficients of both the x and y are the same it is easy. Just add Equations 1 and 2 to eliminate the y variable since the y coefficients are of opposite signs. If, however, the x variable was to be eliminated first, then Equations 1 and 2 would need to be subtracted since the coefficients are of the same signs.

Adding Equation 1 to 2 gives

$$2x = 10 \quad \Rightarrow \quad x = 5$$

Now substituting this $x = 5$ into Equation 1 gives

$$x + y = 6 \Rightarrow \quad 5 + y = 6 \quad \Rightarrow \quad y = 1$$

Finally, the solutions are $x = 5$, $y = 1$.

Note: Remember to check the answers by seeing if they fit both the original Equations 1 and 2.

Example 1.37

Solve the following simultaneous equations:

$$x + 2y = 5 \tag{1}$$

$$3x - y = 1 \tag{2}$$

Now both the coefficients of x and y are different, so the task is first to make one of the variable coefficients the same. Here there is no preferred choice and so try to eliminate the y variable.

To make the coefficients of y the same, multiply Equation 2 by 2, which gives

$$x + 2y = 5 \qquad\qquad (3)$$

$$6x - 2y = 2 \qquad\qquad (4)$$

So now adding Equations 3 and 4 gives

$$7x = 7 \quad \Rightarrow \quad x = 1$$

and substituting this $x = 1$ back into Equation 1 gives

$$x + 2y = 5 \quad \Rightarrow \quad 1 + 2y = 5 \quad \Rightarrow \quad 2y = 4 \quad \Rightarrow \quad y = 2$$

Final solutions are $x = 1$, $y = 2$.

Finally, the most difficult situation for simultaneous equations follows.

Example 1.38

Solve the simultaneous equations

$$2x + 3y = 13 \qquad\qquad (1)$$

$$7x - 5y = -1 \qquad\qquad (2)$$

Solution: Here there is no simple choice to eliminate any of the variables. To eliminate the y variable, Equation 1 gets multiplied by 5 and Equation 2 gets multiplied by 3 and this then has both the y coefficients equal to 15, that is, the lowest common multiple of 3 and 5 as follows:

$$10x + 15y = 65 \qquad\qquad (3)$$

$$21x - 15y = -3 \qquad\qquad (4)$$

Adding Equations 3 and 4 together gives

$$31x = 62 \quad \Rightarrow \quad x = 2$$

substitute this $x = 2$ into Equation 1 gives,

$$2x + 3y = 13 \quad \Rightarrow \quad 4 + 3y = 13 \quad \Rightarrow \quad 3y = 9 \quad \Rightarrow \quad y = 3$$

The final solution is $x = 2$, $y = 3$.

1.5.2   Substitution Method

The second method of solving simultaneous linear equations involves rearranging one of the equations and substituting into the other. This technique is called the method of substitution. It requires making one of the variables the subject and then using this value into the second equation.

Example 1.39

Solve the following simultaneous equations using the method of substitution:

$$7x + 2y = 11 \tag{1}$$

$$4x + y = 7 \tag{2}$$

Solution: Here it is easier (by avoiding fractions) to rearrange Equation 2 to make y the subject to give

$$y = 7 - 4x$$

This value for y can now be substituted into Equation 1 to give

$$7x + 2(7 - 4x) = 11$$
$$7x + 14 - 8x = 11$$
$$-x = -3$$
$$x = 3$$

Using this value for x into $y = 7 - 4x$ gives $y = 7 - 12 \Rightarrow y = -5$. The final solution is then $x = 3$, $y = -5$.

1.6   Quadratic Equations

We have seen that equations like $2x + 1 = 0$ and $x + 2 = 0$ are examples of linear equations. Now, suppose one has an equation like this: $(x + 2)(2x - 1) = 0$. This is not a linear equation; it is called a *quadratic equation* because if the two sets of brackets are expanded, it gives $x^2 + 3x - 2 = 0$.

Note: Here the highest power of the variable x is now 2, and the word quadratic *is derived from the Latin word* quadratus *meaning "square."*

A quadratic equation has not one solution like the linear equations already seen, but two solutions. What are they? Start by looking at the shape of this equation:

$$\text{If (one number)} \times \text{(another number)} = 0$$

The only way that this can happen is if one of the numbers in the brackets is zero. In other words, either $x + 2 = 0$ or $2x - 1 = 0$. Now, it has already been shown how to solve simple linear equations like these.

If $x + 2 = 0$, then $x = -2$. And if $2x - 1 = 0$, then $2x = 1 \Rightarrow x = \dfrac{1}{2}$.

So the answers to the equation $(x + 2)(2x - 1) = 0$ are $x = -2$ or $x = 0.5$.

Example 1.40

One can write the answers to the following quadratic equations fairly easily.

$$(x - 4)(3x + 1) = 0 \qquad \text{Solution: } x = 4 \text{ or } x = -\frac{1}{3}$$

$$(10x + 1)(5x - 2) = 0 \qquad \text{Solution: } x = -\frac{1}{10} \text{ or } x = \frac{2}{5}$$

$$x(2x - 1) = 0 \qquad \text{Solution: } x = 0 \text{ or } x = \frac{1}{2}$$

$$(x - 7)(8x - 3) = 0 \qquad \text{Solution: } x = 7 \text{ or } x = \frac{3}{8}$$

These equations are quadratic equations that have already been *factorized*, that is, written as

$$(\text{one expression}) \times (\text{another expression}) = 0$$

that is, $(x - a)(x - b) = 0$, where a and b are called the "roots" of the equation.

Leaving the topic of factorizing quadratics aside, let's turn to the situation when the quadratic equation doesn't factorize. For example, $x^2 + 8x + 7 = 0$ factorizes and can quickly be solved as shown earlier. But changing the final number to say 8, that is, $x^2 + 8x + 8 = 0$, means that this cannot be done by the factorizing method. Also in realistic engineering problems it is highly unlikely that the quadratic equation generated will factorize. What then? In all general cases, a *quadratic formula* can be used to solve the problem. It is better to consider this method as it can be used in all cases encountered.

1.6.1 Solving Quadratic Equations Using the Formula

Given the general quadratic equation,

$$ax^2 + bx + c = 0 \tag{1.2}$$

where a, b, and c are numbers, then the solutions to this equation are given by the formula known as the general solution to a quadratic equation as

$$x = \frac{-b \pm \sqrt{b^2 - 4ac}}{2a} \tag{1.3}$$

This term underneath the square root sign is given a special name called the *discriminant* and the symbol Δ, where, $\Delta = b^2 - 4ac$.

Notes:

- If the number under the square root sign is *positive*, there are *two real and distinct* solutions.
- If the number under the square root sign is *zero*, there is *one real* and *repeated* solution.
- If the number under the square root sign is *negative*, there are *no real* solutions. This topic is dealt with in Chapter 5, where it will be shown that in this third case the answers can in fact be written as *complex solutions*.

Example 1.41

Solve the equation $x^2 + 8x + 8 = 0$ correct to four decimal places.

Solution: First compare the given quadratic equation with the general quadratic $ax^2 + bx + c = 0$. This then determines the values of the coefficients a, b, and c to be used in the formula as $a = 1$, $b = 8$, and $c = 8$. Substituting these values into the general formula given by Equation 1.3 gives

$$x = \frac{-b \pm \sqrt{b^2 - 4ac}}{2a}$$

$$x = \frac{-8 \pm \sqrt{8^2 - 4(1)(8)}}{2(1)} = \frac{-8 \pm \sqrt{32}}{2} = \frac{-8 \pm 4\sqrt{2}}{2} = -4 \pm 2\sqrt{2}$$

Thus $x = -1.1716$ or $x = -6.8284$.

Example 1.42

Solve the equation $4x^2 - 3x - 11 = 0$ correct to three decimal places.

Solution: Again, first compare the given quadratic equation with the general quadratic $ax^2 + bx + c = 0$. Determining the values of the coefficients a, b, and c as follows: $a = 4$, $b = -3$, and $c = -11$. Substituting these values into the general formula given by Equation 1.3 gives

$$x = \frac{-b \pm \sqrt{b^2 - 4ac}}{2a}$$

$$x = \frac{-(-3) \pm \sqrt{(-3)^2 - 4(4)(-11)}}{2(4)} = \frac{3 \pm \sqrt{185}}{8} = \frac{3 \pm 13.60}{8}$$

Thus, $x = 2.075$ or $x = -1.325$.

Note: All quadratic equations can are solved in the same way with just different coefficients for the values of a, b, *and* c.

1.7 Trigonometry

Many problems in engineering especially mechanics involve forces and these can be represented as either right angled triangles or general scalene triangles. The need to solve these problems requires solutions to triangular problems. First, the methods of solution to right angled triangles is considered.

1.7.1 Right-Angled Triangles

A right-angled triangle is one in which one of the angles is 90 degrees. Consider the general right-angled triangle shown in Figure 1.2. First, the labeling of the sides is very important. The longest side of the right-angled triangle is always called h the hypotenuse, the side facing opposite the angle is o the opposite, and the side next to the angle is called a the adjacent side, as shown in Figure 1.2.

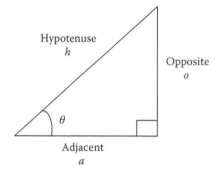

Figure 1.2 A general right-angled triangle.

When dealing with just the sides of the triangle, *Pythagorean theorem* can be used to relate the different sides as follows:

$$h^2 = a^2 + o^2$$

(1.4)

Hence, given any two sides of a right-angled triangle, the third side can be calculated using Equation 1.4.

Example 1.43

Find the missing length AB as shown in the triangle given in Figure 1.3.

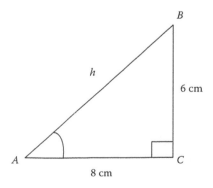

Figure 1.3 A right angled triangle with a missing side.

Solution: Using the labeling notation of Figure 1.2 gives $a = 8$ cm and $b = 6$ cm. Then using Equation 1.4, side h can be calculated as

$$h^2 = a^2 + o^2 = 8^2 + 6^2 = 64 + 36 = 100$$

giving $h = \sqrt{100} = 10$ cm.

When dealing with angles and sides, for the right-angled triangle there are three trigonometric ratios $\sin\theta$, $\cos\theta$, and $\tan\theta$ and these are defined as follows:

$$\sin\theta = \frac{o}{h} \tag{1.5}$$

$$\cos\theta = \frac{a}{h} \tag{1.6}$$

$$\tan\theta = \frac{o}{a} \tag{1.7}$$

These three formulae can more easily be remembered by using the acronym:

$$\text{SOH CAH TOA} \tag{1.8}$$

Note: Given that $\sin\theta = \dfrac{o}{h}$ then to find the angle θ, the inverse sine (i.e., $\sin^{-1}$) function would have been used on both sides to give $\theta = \sin^{-1}\left(\dfrac{o}{h}\right)$. Similarly, inverse function expressions exist for the cosine and tangent functions.

When solving any triangle problem first make sure all sides are labeled correctly, and then identify which one of the ratios is required to solve the problem.

Example 1.44

Find the missing sides o and a in the triangle shown in Figure 1.4.

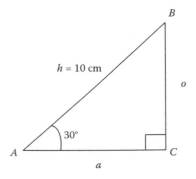

Figure 1.4 Right-angled triangle with unknown sides.

Solution: To find the side o, use a trigonometry identity involving the sides o and side h. Using, SOH CAH TOA, that is, $\sin\theta = \dfrac{o}{h}$ needs to be used and substituting in values for θ and h gives

$$\sin 30° = \frac{o}{10} \quad \Rightarrow \quad o = 10\sin 30° \quad \Rightarrow \quad o = 5\,\text{cm}$$

Example 1.45

Find the angle θ in the triangle given in Figure 1.5.

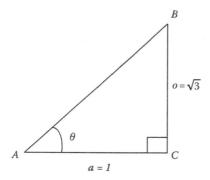

Figure 1.5 Right-angled triangle with missing angle.

Solution: Having the sides o and a, use a trigonometric identity involving sides o and a. SOH CAH TOA leads to the need to use $\tan\theta = \dfrac{o}{a}$. Substituting in values for a and o gives

$$\therefore \quad \tan\theta = \frac{\sqrt{3}}{1} \quad \Rightarrow \quad \theta = \tan^{-1}\left(\frac{\sqrt{3}}{1}\right),$$

that is, $\theta = 60°$.

1.7.2   Scalene Triangles (Sine and Cosine Rules)

Many of the problems encountered will form a general triangle and this is called a scalene triangle. In such cases, two new formulae are necessary to solve these triangular problems. The first rule is called the *sine rule* and the second rule is called the *cosine rule*.

1.7.2.1   Sine Rule

Given a general triangle as shown in Figure 1.6, the labeling is again very important as the formulae depend on it. Here, the angles are denoted by the capital letters and the side opposite that angle is denoted by the corresponding lowercase letter.

The sine rule relates the sine of an angle with its opposite side as follows:

$$\frac{a}{\sin A} = \frac{b}{\sin B} = \frac{c}{\sin C} \tag{1.9}$$

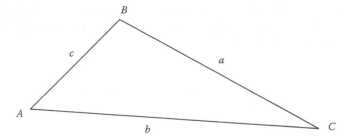

Figure 1.6 General scalene triangle.

Note: In Equation 1.9 *any two of the relationships are used as required, that is,*

$$\frac{a}{\sin A} = \frac{b}{\sin B} \quad \text{or} \quad \frac{a}{\sin A} = \frac{c}{\sin C} \quad \text{or} \quad \frac{b}{\sin B} = \frac{c}{\sin C}$$

Example 1.46

In the triangle given in Figure 1.7, find the length of side AC.

Solution: First find the missing angle B, angle B is given by $180° - (83° + 62°) = 35°$

So now using the sine rule, $\dfrac{a}{\sin A} = \dfrac{b}{\sin B} = \dfrac{c}{\sin C}$, since the side AC (i.e., side b) is required and side a and angle A are also known, then using

$$\frac{a}{\sin A} = \frac{b}{\sin B} \quad \Rightarrow \quad \frac{10}{\sin 83} = \frac{b}{\sin 35} \quad \Rightarrow \quad b = \frac{10 \times \sin 35}{\sin 83} = 5.78\,\text{cm}$$

Note: It is not always possible to make use of the sine rule directly, so sometimes there is a need for the second rule.

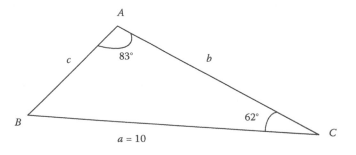

Figure 1.7 A general triangle to find missing sides and angles.

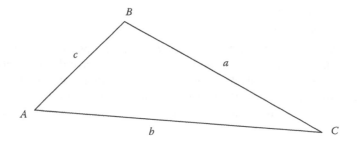

Figure 1.8 General scalene triangle.

1.7.2.2 Cosine Rule

The cosine rule is needed when there are two given sides and only the angle between them is given. Consider the general triangle in Figure 1.8.
The Cosine rule is stated as

$$a^2 = b^2 + c^2 - 2bc \cos A \tag{1.10}$$

And by symmetry,

$$b^2 = a^2 + c^2 - 2ac \cos B \tag{1.11}$$

$$c^2 = a^2 + b^2 - 2ab \cos C \tag{1.12}$$

To find the angles when all three sides are known, Equation 1.10 can be rearranged to give

$$\cos A = \frac{b^2 + c^2 - a^2}{2bc} \tag{1.13}$$

Example 1.47

Find the length AC given in Figure 1.9.

Solution: Using the cosine rule given by Equation 1.11, $b^2 = a^2 + c^2 - 2ac \cos B$ gives

$$b^2 = 10^2 + 8^2 - 2 \times 10 \times 8 \times \cos 60° \implies b^2 = 84 \implies b = 9.17 \, \text{cm}$$

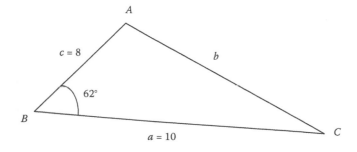

Figure 1.9 A general triangle using the cosine rule.

1.7.3   Resultant Forces

1.7.3.1   Adding Two Forces

Consider two forces of magnitude, F_1 and F_2, that act upon a particle. If these forces are placed end to end, it can be seen that they have the same effect as a single force of magnitude F, as in Figure 1.10. This force is known as the *resultant force*. The resultant force will form the third side in a triangle of forces. To calculate this resultant force and its direction, use of the cosine and sine rule are made.

Figure 1.10 Resultant forces forming general triangles.

Example 1.48

Two forces of magnitude, 6 N and 5 N, act on a particle. The angle between the forces is 40°. Find the magnitude and direction of the resultant force.

Solution: Drawing the resultant diagram as shown in Figure 1.11.
First, using the cosine rule in the triangle of forces to calculate the size of the resultant force gives

$$F^2 = 6^2 + 5^2 - 2 \times 6 \times 5 \cos 140° = 106.96$$
$$F = 10.34\,\text{N}$$

To find the direction of the resultant force with respect to the 6 N force means finding the angle θ in Figure 1.11. This can be done using the sine rule as

$$\frac{5}{\sin\theta} = \frac{F}{\sin 140} \quad \Rightarrow \quad \sin\theta = \frac{5 \times \sin 140}{10.34} \quad \Rightarrow \quad \theta = 18.1°$$

So, the resultant force has magnitude 10.34 N and is at an angle 18.1° to the 6 N force.

Figure 1.11 Diagram showing resultant force forming a general triangle of forces.

1.7.4   Basic Trigonometric Identities

Having already seen the trigonometric ratios for $\sin \theta$, $\cos \theta$, and $\tan \theta$, that is,

$$\sin \theta = \frac{o}{h}, \quad \cos \theta = \frac{a}{h}, \quad \tan \theta = \frac{o}{a}$$

There are important trigonometric identities that can be derived and are very useful in applications later on. Two basic identities are

$$\cos^2 \theta + \sin^2 \theta = 1 \tag{1.14}$$

$$\tan \theta = \frac{\sin \theta}{\cos \theta} \tag{1.15}$$

Note: These can be proved from the aforementioned trigonometric ratios and can be used to prove other trigonometric identities as well as solve trigonometric equations.

Example 1.49

Find the value of $\tan \theta$ when $\sin \theta = \dfrac{3}{5}$ and $\cos \theta = -\dfrac{4}{5}$.

Solution:

$$\tan \theta = \frac{\sin \theta}{\cos \theta} = \frac{3}{5} \div \left(-\frac{4}{5} \right) = \frac{3}{5} \times \left(-\frac{5}{4} \right) = -\frac{3}{4}$$

Example 1.50

Show that $\dfrac{1-\sin \theta}{\cos \theta} \equiv \dfrac{1}{\cos \theta} - \tan \theta.$

Note: When asked to "show that" or "prove that" then consider starting with one side of the identity and, step by step, reduce it to the same form as the other side of the identity.

Solution:

$$\text{LHS:} \quad \frac{1-\sin \theta}{\cos \theta} \equiv \frac{1}{\cos \theta} - \frac{\sin \theta}{\cos \theta} \equiv \frac{1}{\cos \theta} - \tan \theta$$

Example 1.51

Show that $\dfrac{2-\cos^2 \theta}{1+\sin^2 \theta} \equiv 1.$

Solution: Since $\cos^2\theta + \sin^2\theta = 1 \quad \Rightarrow \quad \cos^2\theta = 1 - \sin^2\theta$
So,

$$\text{LHS: } \frac{2-\cos^2\theta}{1+\sin^2\theta} \equiv \frac{2-1(1-\sin^2\theta)}{1+\sin^2\theta} \equiv \frac{1+\sin^2\theta}{1+\sin^2\theta} \equiv 1$$

There are other useful compound formulae for the sine, cosine, and tangent as follows:

$$\cos(A \pm B) = \cos A \cos B \mp \sin A \sin B \tag{1.16}$$

$$\sin(A \pm B) = \sin A \cos B \pm \cos A \sin B \tag{1.17}$$

$$\tan(A \pm B) = \frac{\tan A \pm \tan B}{1 \mp \tan A \tan B} \tag{1.18}$$

Other important trigonometric formulae can be derived from Equations 1.16 and 1.17 by adding and subtracting these to give the following:

$$\cos(A)\cos(B) = \frac{1}{2}\big(\cos(A+B) + \cos(A-B)\big) \tag{1.19}$$

$$\sin(A)\sin(B) = \frac{1}{2}\big(\cos(A-B) - \cos(A+B)\big) \tag{1.20}$$

$$\sin(A)\cos(B) = \frac{1}{2}\big(\sin(A+B) + \sin(A-B)\big) \tag{1.21}$$

The preceding formulae are very important whenever the need arises to transform the product of sines and cosines into sums and are a very useful in the techniques of integration and applications in areas such as Fourier series, seen later in Chapter 9.

1.7.5 Radian Measure

When it comes to measuring angles generally the measurement that has been used is degrees. A whole circle is 360°, a straight line is 180°, and a right angle is 90°, for example. However, when it comes to finding gradients of curves, or rates of change (i.e., to differentiate trigonometric functions), a different unit of measurement is used called the radian.

Radian is short for the *radius angle*, and it means the angle given at the center of a circle by an arc of one radius as shown in Figure 1.12.

Now the relationship between radians and degrees can be determined as follows: There are 360° in a complete revolution of the circle, the circumference of a circle is $2\pi r$, and so the number of radiuses r around the circumference is given by $\frac{2\pi r}{r} = 2\pi$ radians. Therefore, this gives an important formula relating degrees to radians:

$$360° \equiv 2\pi \text{ radians} \tag{1.22}$$

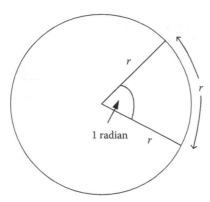

Figure 1.12 Definition of the radian angle.

Table 1.2 Relationships between Degrees and Radians

Angle in Degrees	Angle in Radians
360°	2π
180°	π
90°	$\dfrac{\pi}{2}$
60°	$\dfrac{\pi}{3}$
45°	$\dfrac{\pi}{4}$

The standard relationships that exist between degrees and radians is shown in Table 1.2.

When talking about an angle in degrees, one should write the degrees sign as 45.2°. When talking about an angle in radians, a sign is not used. Thus $\cos(60°) = \cos\left(\dfrac{\pi}{3}\right) = 0.5$ and sin (25°) = 0.4226, but sin (25) = −0.1324 because this means sin (25 radians).

1.7.5.1   Radians on the Calculator

Trigonometric problems can be solved in either degrees or radians. Probably the need to use radians will arise when doing certain types of problems involving calculus. The calculator can be set to work in radians instead of degrees using the Mode button.

1.8   Statistics

1.8.1   Introduction

Statistics deals with all aspects of data: collecting the data, pictorial representation of the data, and numerical analysis and drawing final conclusions on findings. There are vast areas where statistics are used ranging from opinion pollsters (testing public opinions on issues), governmental national statistics, and in science and engineering testing theories.

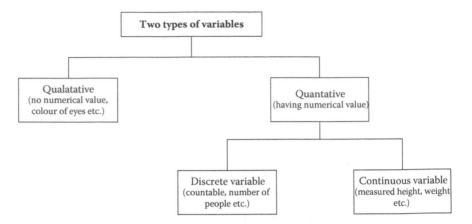

Figure 1.13 Different types of variables in statistics.

There are two types of variables, as shown on Figure 1.13, and any quantity that varies is called a variable.

There are some different aspects to statistical data that need defining and further explanation:

Primary data—This is data collected for a specific investigation. This could involve carrying out actual experiments, sending out questionnaires, and so forth.

Secondary data—Organizations and governments publish a vast amount data on a wide variety of subjects that appear in different publications. They provide useful data for investigations but they were not collected specifically for the investigation.

Population—This is all the possible data for the research question. If the question was to determine the average height of male adults in the United Kingdom, then the population would need to include measuring every single adult male.

Sample—This is just part of the data set. In some practical situations, it makes no sense to consider the whole population, as this may be too large (i.e., in millions) and so a smaller sample is considered more appropriate. Sampling is very important and useful; it reduces the amount of data that needs to be collected.

Sampling without bias—A bias is anything that makes the sample unrepresentative (e.g., asking only certain members of a community their views a topic).

Random sampling—A random sample of size n is a sample selected so that all possible samples of size n have an equal chance of being selected.

1.8.2   Measures of Averages

There are three main measures of *average* of a set of numerical data:

Mode—The one that occurs most frequently (or often).

Median—Arrange the data in order of magnitude (size), then find the central value. If there is an even number of data values, then use halfway between the two middle values.

Mean—Add all the data values and divide by the total number of data points.

Note: For most purposes the mean is considered the most useful measure of average since it uses all data.

The symbol μ (pronounced myü) is used for the mean of the whole population and the symbol $\bar{x}$ (x bar) is used for the mean of the sample. The definition of the mean can be more concisely written in formula form as

$$\bar{x} = \frac{\sum x_i}{n} \tag{1.23}$$

Note: Here the symbol $\sum x_i$ means to sum or add together all the x_i values.

Equation 1.23 is used for a simple set of data (i.e., a small number of data points) and where n is the number of data points.

Example 1.52

The number of fires reported in 20 consecutive weeks is given by the data set as:

4	7	12	13	0	5	21	13	10	6
6	8	15	9	6	0	14	12	6	8

Find the modal, median, and mean number of fires per week.

Solution: It is first easier to put the data set in order as

0 0 4 5 6 6 6 6 7 8 8 9 10 12 12 13 13 14 15 21

The modal number is, 6 as this occurs more times than all others.

For the median, since there is an even number of data points (i.e., 20 data points) look at the 10th and 11th data values. In this case they are both 8 and 8, which gives the middle of these as

$$\text{Median} = \frac{8+8}{2} = 8$$

For the mean, use the formula given by Equation 1.23:

$$\bar{x} = \frac{\sum x_i}{n} = \frac{175}{20} = 8.75.$$

1.8.2.1   Data in a Frequency Table

Sometimes when large data sets are involved, it is no longer easy or practical to write out the data points as a list especially if there are hundreds or thousands of data points. In this situation, it is more convenient to use a frequency table to represent the data values. The *frequency* is the number of times an observation occurs. The formula for the mean now modifies as

$$\bar{x} = \frac{\sum x_i f_i}{\sum f_i} \tag{1.24}$$

where f_i is just the frequency associated with the ith data point.

Example 1.53

A survey was carried out to see the distribution of the ages (in years) of fire engines in a particular region of the country. Twenty-seven vehicles were surveyed and the results are shown in Table 1.3. Find the mode, median, and mean age of the fire engines.

Solution: The first thing to do is redraw the table with an extra column for the product term $x_i f_i$, as shown in Table 1.4. The columns can be filled in and the products computed. At the bottom of the columns, the sums of the columns can also be calculated.

Mode: 0 (i.e., highest frequency). These will be new fire engines.

Median: There are 27 data points. To find the middle, $\frac{27+1}{2} = 14$th $\Rightarrow$ data point $\Rightarrow$ 2 years.

Mean: $\bar{x} = \dfrac{\sum x_i f_i}{\sum f_i} = \dfrac{45}{27} = 1.67$ years.

Table 1.3 A Frequency Table
Showing Ages of Fire Engines

Ages of Fire Engine, x_i	Frequency, f_i
0	9
1	4
2	6
3	5
4	2
5	0
6	1

Table 1.4 Constructing the Extended Frequency Table

Age of Fire Engine, x_i	Frequency, f_i	Product, x_if_i
0	9	$0 \times 9 = 0$
1	4	$1 \times 4 = 4$
2	6	$2 \times 6 = 12$
3	5	$3 \times 5 = 15$
4	2	$4 \times 2 = 8$
5	0	$5 \times 0 = 0$
6	1	$6 \times 1 = 6$
	$\sum f_i = 27$	$\sum x_if_i = 45$

1.8.2.2 Grouped Data

Sometimes the individual data points are not provided but only the grouped frequency table is given. In this case it is not possible to give the exact values of the three averages but only estimates.

For the mean, the method is the same as for Example 1.48, but first the midpoint of the grouped data set must be determined. This is a good approximation since some of the data values will be higher than this and some will be lower than this middle value. The next example shows how this method works.

Example 1.54

Consider the marks for 61 students who took a test as given in Table 1.5. Find the modal class, median, and the mean percentage mark for this class.

Solution:

Modal class: 51%–60% (this is the most frequently occurring).
Median: Middle value, $\frac{27+1}{2} = 31$st data value, so it's the 6th data value

in the class interval 51–60 and so needs calculating as follows:

$$51 + \frac{6}{22} \times (\text{class width} = 10) = 53.73\%$$

Table 1.5 A Frequency Table Showing Grouped Data

Marks%, x_i	Frequency, f_i
1–10	1
11–20	4
21–30	3
31–40	7
41–50	10
51–60	22
61–70	8
71–80	3
81–90	2
91–100	1

Table 1.6 Calculating an Estimate of the Mean for a Grouped Data Set

Marks%, x_i	Frequency, f_i	Midpoint%, x_m	Product, $x_m f_i$
1–10	1	5.5	5.5
11–20	4	15.5	62
21–30	3	25.5	76.5
31–40	7	35.5	248.5
41–50	10	45.5	455
51–60	22	55.5	1221
61–70	8	65.5	524
71–80	3	75.5	226.5
81–90	2	85.5	171
91–100	1	95.5	95.5
	$\sum f_i = 61$		$\sum x_m f_i = 3085.5$

The mean is obtained as before, but now the midpoint of the class width is needed since the exact data values are not known, as shown in Table 1.6. So the estimated mean is

$$\frac{\sum x_m f_i}{\sum f_i} = 50.58\%$$

1.8.3 Measures of Spread

In the previous section it was seen that for a data set an average value could be calculated to indicate something about the data points. However, this does not say anything about how the data points are spread out. There are three common measures of spread: The range (R), the interquartile range (IQR), and the standard deviation (σ) are defined next.

1.8.3.1 Range

The *range* for a set of data is the difference between the highest value and the lowest value (i.e., it considers the extreme values of the data set).

Example 1.55

See the following two data sets A and B:

A: 4, 5, 5, 6, 7, 9
B: 1, 3, 3, 5, 6, 8, 10, 12

The range for set A is R = 9 – 4 = 5 and for data set B is R = 12 – 1 = 11. So it can be seen that even though the two data sets have the same mean (i.e., 6), the set B has more variation in the data points than set A.

1.8.3.2 Interquartile Range

When a set of *n* data is written in order of magnitude (size), the median is given by $\frac{(n+1)}{2}$ th item of data.

The *quartiles* are found in a similar way:

- The lower quartile is the median of the lower half of the data.

- The upper quartile is the median of the upper half of the data.

- Then the Interquartile range (IQR) = Upper quartile – Lower quartile.

Example 1.56

Find the interquartile range (IQR) for the following set of data:

$$24, \ 24, \ 25, \ 26, \ 26, \ 26, \ 27, \ 27, \ 30, \ 33, \ 33, \ 35, \ 35, \ 36, \ 43$$

Solution: There are 15 data values, so the median is the eighth data value. Thus, median = 27.

Lower quartile has 7 data values $\Rightarrow$ median of this is the fourth value = 26.

Upper quartile has 7 data values $\Rightarrow$ median of this is 35.

$$\therefore \ \text{IQR} = \text{UQ} - \text{LQ} = 35 - 26 = 9$$

1.8.3.3 Standard Deviation (σ)

The range and IQR do not use all the data to measure the spread. A more useful indicator of the average spread of data about the mean is the *standard deviation* and is used along with the mean widely in science and engineering. The formulae for standard deviation for a population and for a sample are as follows.

1.8.3.3.1 *Population Standard Deviation (σ)*

Standard deviation of a population for a simple data set $x_1, x_2, x_3, \ldots, x_n$, is denoted by σ and is given by

$$\sigma = \sqrt{\frac{\sum (x_i - \mu)^2}{n}} \tag{1.25}$$

where μ is the population mean and n is the number of data points.

For a frequency distribution

$$\sigma = \sqrt{\frac{\sum (x_i - \mu)^2 f_i}{\sum f_i}} \tag{1.26}$$

where f_i are the individual frequencies associated with the x_i.

Note: The units of σ are the same as the units of x and $\bar{x}$.

1.8.3.4   Sample Standard Deviation (s)

If the data is from a small sample of a population, the population mean μ is not calculated but rather the sample mean $\bar{x}$. Then the estimated standard deviation (s) is given by

$$s = \sqrt{\frac{\sum (x_i - \bar{x})^2}{n-1}} \qquad (1.27)$$

Here, division is by $n-1$ instead of n.

Again, for a frequency distribution

$$s = \sqrt{\frac{\sum (x_i - \bar{x})^2 f_i}{\sum f_i - 1}} \qquad (1.28)$$

Note: Generally, it is much easier to calculate the mean and standard deviation using a calculator (in statistics mode).

Example 1.57

Look at the following set of data and find the mean and standard deviation:

$$x_i: 1, 2, 3, 4, 5$$

Solution:

Mean: $\mu = \dfrac{\sum x_i}{n} = \dfrac{15}{5} = 3$

Now use Table 1.7 to find the standard deviation σ as

$$\therefore \ \sigma = \sqrt{\frac{\sum (x_i - \mu)^2}{n}} = \sqrt{\frac{10}{5}} = \sqrt{2} = 1.414$$

Table 1.7 Calculating the Standard Deviation of a Simple Data Set

x_i	$(x_i - \mu)$	$(x_i - \mu)^2$
1	−2	4
2	−1	1
3	0	0
4	1	1
5	2	4
		$\sum (x_i - \mu)^2 = 10$

Example 1.58

The number of fires over 52 weeks was recorded at a particular fire station. The results are shown as follows:

No. of fires, x_i	4	5	6	7	8	9	10
Frequency, f_i	3	10	18	6	5	6	4

Calculate the mean and standard deviation of the number of fires per week.

Solution: The calculation are shown in Table 1.8 as

$$\mu = \frac{\sum x_i f_i}{\sum f_i} = \frac{346}{52} = 6.65$$

$$\sigma = \sqrt{\frac{\sum (x_i - \mu)^2 f_i}{\sum f_i}} = \sqrt{\frac{143.77}{52}} = \sqrt{2.765} = 1.66$$

Finally, the square of the standard deviation (σ^2) is given a special name as the *variance*.

Table 1.8 Calculating Mean and Standard Deviation for Data in a Frequency Table

x_i	f_i	$x_i f_i$	$(x_i - \mu)$	$(x_i - \mu)^2 f_i$
4	3	12	−2.65	21.0675
5	10	50	−1.65	27.225
6	18	108	−0.65	7.605
7	6	42	0.35	0.735
8	5	40	1.35	9.1125
9	6	54	2.35	33.135
10	4	40	3.35	44.89
	$\sum f_i = 52$	$\sum x_i f_i = 346$		$\sum (x_i - \mu)^2 f_i = 143.77$

1.8.4 Change of Scale

The weekly wages of employees in a small company have a mean of £290 and standard deviation of £42. If there is a pay rise of £15 for each employee, what happens to the mean and standard deviation? Since each wage is increased by £15, the mean wage will be increased by £15 to £305. However, the standard deviation measures variability and this is unchanged since all wages have increased by the same amount. If instead of a flat rate increase an increase of 10% is given to each employee, the variability would increase because the higher paid employees would get a larger rise. In this case both the mean and standard deviation would increase by 10%. In summary, if a variable is increased by a constant amount, its average will be increased by this amount, but the spread will be unchanged. If the variable is multiplied by a constant amount, both its average and spread will be multiplied by this amount.

1.9   Applications

Example 1.59: Relationship between Sizes of Pool Fires to Flame Heights

A schematic diagram of a pool fire with its flame height is shown in Figure 1.14. From pool fire experiments, a power law fit was determined relating the heat release rate $\dot{Q}$ (kW) and the diameter of the pool fire D (m) to the flame height L (m) using Heskestad's equation as follows:

$$L = 0.235\dot{Q}^{\frac{2}{5}} - 1.02D \tag{1.29}$$

Knowing the heat release rate for a fire and its diameter size, then using Equation 1.29 can predict the height of the flame.

Consider a 500 kW diesel pool fire with a diameter of 1.5 m, the flame height is given by substituting in the data values into Equation 1.29 as

$$L = 0.235(500)^{\frac{2}{5}} - 1.02(1.5) = 1.29 \text{ m}$$

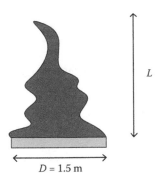

L

$D = 1.5$ m

Figure 1.14 Diesel pool fire with its associated flame height.

Example 1.60: Enclosure Fire with a Ceiling

Figure 1.15 shows the development of a fire within an enclosure of height H. When the fire plume interacts with the ceiling it cannot travel further upward, so it creates a horizontal ceiling jet. The temperature of the

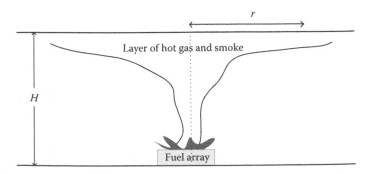

r

Layer of hot gas and smoke

H

Fuel array

Figure 1.15 Smoke layer development with an enclosure ceiling.

smoke layer measured radially far away from the fire is given by Alpert's equation as

$$T_{jet} - T_0 = 5.38 \frac{(\dot{Q}/r)^{\frac{2}{3}}}{H} \qquad (1.30)$$

where T_{jet} is the smoke layer temperature, T_0 is the ambient room temperature, $\dot{Q}$ is the heat release rate, r is the radial distance of the smoke, and H is the room height.

If a sprinkler was located 5m radially from the fire with a ceiling height of 3m and the fire size is 1200 kW with an ambient temperature of 20°C then what is the expected gas temperature at the sprinkler?

Equation 1.30 can be transposed to make T_{jet} the subject of the formula to give

$$T_{jet} = T_0 + 5.38 \frac{(\dot{Q}/r)^{\frac{2}{3}}}{H} \qquad (1.31)$$

Substituting in the values of the parameters into Equation 1.31 gives

$$T_{jet} = 20 + 5.38 \frac{(240)^{\frac{2}{3}}}{3} = 89°C$$

Note: If the temperature to activate a sprinkler or smoke detector is known, then the equation can be transposed to solve for $\dot{Q}$. This can be useful for design purposes in fire calculations to determine the fuel size required to activate a sprinkler or smoke detector.

Example 1.61: Direction of Smoke Flow with Wind Effects

A smoking fire in a room is located at a point A as shown in Figure 1.16. Smoke flows vertically upward with a velocity $V_s = 1.2$ ms⁻¹. If a window is open and the wind with velocity $V_w = 0.6$ ms⁻¹ is flowing due east, calculate

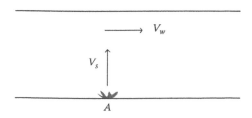

Figure 1.16 Smoke movement with wind effects.

the *resultant velocity* V_R of the smoke flow and the direction it flows with reference to the horizontal floor.

First constructing a triangle of velocities gives Figure 1.17.

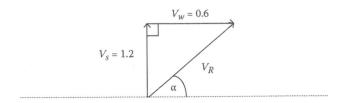

Figure 1.17 Resultant velocity of the smoke flow.

It can be seen from the right-angled triangle that $V_R^2 = 1.2^2 + 0.6^2 = 18$ ∴ $V_R = 1.34\,\text{ms}^{-1}$.

The direction of the smoke flow is given by the angle α as

$$\tan \alpha = \frac{1.2}{0.6} \Rightarrow \alpha = 63.4°.$$

Example 1.62: Change in Mean and Standard Deviation

A sprinkler, designed to extinguish house fires, is activated at high temperatures. A batch of sprinklers is tested and found to be activated at a mean temperature of 72°C with a standard deviation of 3°C. Find the mean and standard deviation of the temperature in degrees Fahrenheit.

The formula relating degrees centigrade to degrees Fahrenheit is F = 1.8C + 32. So the mean would be multiplied by 1.8 and add to it 32 to give

$$\text{Mean} = (72 \times 1.8) + 32 = 161.6°\text{F}$$

But for the standard deviation, only carry out the multiplication since the addition of 32 will have no effect on the variability giving

$$\text{Standard deviation} = 3 \times 1.8 = 5.4°\text{F}$$

Problems

1.1 Transpose the following formulae for the variable given:

    a. $L = \dfrac{3.6V}{A}$ for A

    b. $F = G + 7H$ for H

    c. $\alpha = \dfrac{k}{\rho c}$ for k

    d. $v = \sqrt{h - gt}$ for t

e. $t = \dfrac{d^2}{4\alpha}$ for d

f. $I = \varepsilon \sigma A T^4$ for T

g. $\tau = t_I \sqrt{\dfrac{g}{H} \dfrac{H^2}{S}}$ for t

h. $E = \dfrac{1}{2} m v^2$ for v

i. $g = \dfrac{h+3}{h-5}$ for h

j. $U = 0.96 \left(\dfrac{\dot{Q}}{H} \right)^{\frac{1}{3}}$ for $\dot{Q}$

1.2 The equation for the smoke jet T_j is given by Equation 1.31 as

$$T_j = T_0 + 5.38 \frac{\left(\dot{Q}/r \right)^{\frac{2}{3}}}{H}$$

   a. By transposing, find an expression for the heat release rate $\dot{Q}$.

   b. Given that $H = 15$m, $r = 8$m, $T_0 = 20°C$, and $T_j = 100°C$ determine the value of $\dot{Q}$.

1.3 Solve the following equations:

   a. $a + b = 2$ and $5a + b = 14$

   b. $2x + 3y = 4.5$ and $5x - 2y = 6.5$

   c. $x^2 + 5x - 3 = 0$

   d. $3x^2 - 6x - 11 = 0$

1.4 A fireboat in Hong Kong starting at a point O is crossing a river that has two parallel banks. The width of the river is 200 m. The water in the river is flowing at a speed of V ms^{-1}. Point A is directly opposite point O on the other bank. The velocity of the boat relative to the water is 20 ms^{-1} at angle of 70° to the bank. The boat lands at a point B were a building is burning, which is 30 m from A. The angle between the actual path of the boat and the bank is $\theta°$. Figure 1.18 shows the corresponding river and the velocities of the boat and the water.

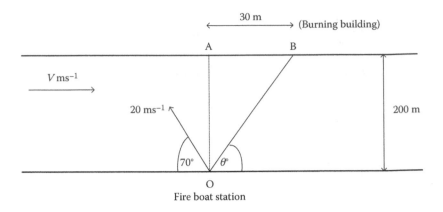

Figure 1.18 Fireboat crossing a river bank.

    a. Find the time it takes the fireboat to cross the river and the actual direction of the boat θ.

    b. Determine the velocity of the water flow.

1.5 The number of defective components reported to a company per week is given as follows:

No. of defective components, x_i	5	7	10	13	16	19
Frequency, f_i	2	3	8	5	3	1

    Calculate the mean number of defective components reported per week and the standard deviation.

2 Introduction to Probability Theory

2.1 Introduction

Most people deal with uncertain events each day in their lives but rarely pause to calculate chances or probabilities. Yet whole professions, such as the insurance industry, pensions, investment advisors, and bookmakers are founded upon probability. In engineering, the ideas of probability are very important when considering the overall reliability of systems and in the areas of risk assessment.

How is the probability of something happening defined? The usual definition supposes that one repeats an "experiment" many times and records the outcomes. For example, roll a fair dice many times and record how often the result is a six.

The probability of a six occurring is given by

$$P(6) = \frac{\text{Number of times a six occurs}}{\text{Total number of throws}}$$

If the die was fair and rolled lots of times, the expected probability would be $\frac{1}{6}$.

Therefore, generally the probability of an event A is defined as

$$P(A) = \frac{\text{Number of ways } A \text{ can occur}}{\text{Total number of possible outcomes}} \qquad (2.1)$$

Note: If A denotes the event A happening, then A' or (A^c) is called the complement of A and denotes the event A not happening.

There are several consequences of this definition. *A probability must lie between zero and 1.* Also, adding the probabilities of all the outcomes, the *total must be 1*. This gives the following: $P(A) + P(A') = 1$ and so

$$P(A') = 1 - P(A) \tag{2.2}$$

For example, for a fair dice $P(1) = P(2) = P(3) = P(4) = P(5) = P(6)$. And of course $\frac{1}{6} + \frac{1}{6} + \frac{1}{6} + \frac{1}{6} + \frac{1}{6} + \frac{1}{6} = 1$. But it can also be concluded, for example, since $P(6) = \frac{1}{6}$, then $P(\text{not a } 6) = 1 - \frac{1}{6} = \frac{5}{6}$.

It is not always possible to assign a probability using equally likely outcomes. If someone wanted to know the probability of going to the bus stop and having to wait for a bus for more than 5 minutes, then some trials would have to be done to find the waiting times. The *relative frequency* of an event is the proportion of times it has been observed to happen. If somebody went for a bus on 40 weekday mornings and on 16 of these they had to wait more than 5 minutes, one could assign a probability of $\frac{16}{40} = 0.4$ to the event of having to wait more than 5 minutes.

2.1.1   Mutually Exclusive Events

Mutually exclusive events are those that cannot both happen, for example, scoring a 3 and scoring a 4 when throwing a dice. One can find the probability that either one event *or* the other happens by adding the probabilities of the two events.

Example 2.1

When rolling a dice, the

$$P(\text{either getting a 3 or 4}) = \frac{1}{6} + \frac{1}{6} = \frac{1}{3}$$

Example 2.2

When choosing one card from a pack, the probability of getting a ten card or a jack card is

$$P(\text{a ten or a jack}) = P(\text{ten}) + P(\text{jack}) = \frac{4}{52} + \frac{4}{52} = \frac{8}{52} = \frac{2}{13}$$

Generally, for *mutually exclusive* events the relationship is

$$P(\text{A or B}) = P(A) + P(B) \tag{2.3}$$

This can be extended for any number of mutually exclusive events as

$$P\left(A_1 \text{ or } A_2 \text{ or } A_3 \ldots\ldots \text{or } A_n \right) = P(A_1) + P(A_2) + P(A_3) + \ldots\ldots + P(A_n)$$

Or in concise form as,

$$P\left[\bigcup_{i=1}^{n} A_i\right] = \sum_{i=1}^{n} P(A_i)$$

If the events are not mutually exclusive these formulae are modified. For example Equation 2.3 now becomes

$$P(A \text{ or } B) = P(A) + P(B) - P(A \text{ and } B) \qquad (2.4)$$

This is because these events can happen together and the probability of B is included in the probability of A and vice versa, so this means that the probability has been included twice and so one of these must be subtracted from the answer.

2.1.2 Independent Events

Independent events are those that have no influence on each other. When tossing a coin twice, the result of the first throw has no effect on the second.

In such a case, the probability of both events happening is found by multiplying the separate probabilities.

Example 2.3

$$P \text{ (head on both tosses of a coin)} = P \text{ (head)} \times P \text{ (head)} = \frac{1}{2} \times \frac{1}{2} = \frac{1}{4}$$

$$P \text{ (rolling two sixes with a dice)} = P \text{ (6)} \times P \text{ (6)} = \frac{1}{6} \times \frac{1}{6} = \frac{1}{36}$$

Generally, for *independent events*:

$$P(A \text{ and } B) = P(A) \times P(B) \qquad (2.5)$$

This can be extended for any number of independent events as

$$P\left(A_1 \text{ and } A_2 \text{ and } A_3 \ldots \ldots \text{and } A_n\right) = P(A_1)P(A_2)P(A_3)\ldots\ldots P(A_n)$$

or in concise form as,

$$P\left[\bigcap_{i=1}^{n} A_i\right] = \prod_{i=1}^{n} P(A_i)$$

If the events are not independent, then one has *conditional probabilities* and these formulae are modified (see later).

Example 2.4

The probability that telephone calls to a railway timetable inquiry service are answered is 0.7. If three calls are made find the probability that all three are answered and exactly two are answered.

Solution: If A is the event of a call being answered, $P(A) = 0.7$, then A' is the probability of a call not being answered, and $P(A') = 1 - 0.7 = 0.3$.

Using the multiplication rule the probability of $AAA = 0.7 \times 0.7 \times 0.7 = 0.343$.

If one call is unanswered it could be the first, second, or the third call:

$A'AA$ with probability $0.3 \times 0.7 \times 0.7 = 0.147$
$AA'A$ with probability $0.7 \times 0.3 \times 0.7 = 0.147$
AAA' with probability $0.7 \times 0.7 \times 0.3 = 0.147$

These three outcomes are mutually exclusive and so one can apply the addition law and find the probability of exactly two calls being answered to be $0.147 + 0.147 + 0.147 = 0.441$.

2.1.3   Conditional Probability

Sometimes the events A and B are not independent, and so the probability of the event A happening depends on the event B happening. This is represented as follows:

$P(A|B)$ denotes the probability that event A happens given that event B happens.

Two events A and B are independent if $P(A) = P(A|B)$.
So the multiplication law given by Equation 2.5 becomes more general as

$$P(\text{A and B}) = P(A) \times P\left(B|A\right) \qquad (2.6)$$

Example 2.5

James buys ten apparently identical oranges. Unknown to him the flesh of two of these oranges is rotten. He selects two of the ten oranges at random and gives them to his friend. Find the probability that

a. Both oranges are rotten.
b. Exactly one of the oranges is rotten.

Solution: Sometimes in probability type problems it is easier to see what is happening using a *tree diagram* (discussed later in Section 2.1.5) as shown in Figure 2.1.

a. The probability of both oranges being rotten is $\dfrac{2}{90} = 0.022$.

b. The probability of exactly one orange being rotten is $\dfrac{16}{90} + \dfrac{16}{90} = 0.35$.

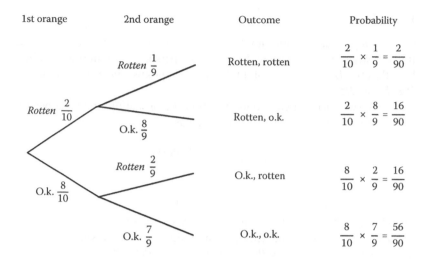

| 1st orange | 2nd orange | Outcome | Probability |

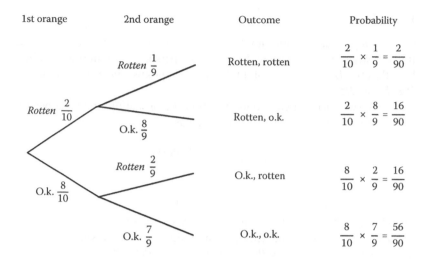

Figure 2.1 Tree diagram showing possible outcomes and probabilities.

2.1.4 Bayes' Theorem

Founded by the Rev. Thomas Bayes (1701–1761), who apart from being a minister was a statistician and philosopher, Bayes' theorem is a fact about probabilities and has a lot of real-world applications. It gives a way of working out what the conditional probabilities should be.

A notation that is used for the following case is that the probability of a hypothesis, H, given that a new piece of evidence E is written as $P(H\backslash E)$.

Bayes' theorem states

$$P(H\backslash E) = \frac{P(E\backslash H)\,P(H)}{P(E)} \tag{2.7}$$

Proof: Generally, we have from the multiplication laws of probabilities:

$$P(H \text{ and } E) = P(H)P(E\backslash H)$$

$$P(E \text{ and } H) = P(E)P(H\backslash E)$$

Now these two are equal, so

$$P(E)\,P(H\backslash E) = P(E\backslash H)\,P(H)$$

$$P(H\backslash E) = \frac{P(E\backslash H)\,P(H)}{P(E)}$$

which is Bayes' theorem.

Example 2.6

One day a person who does not feel well decides to go onto the Internet to find out what might be wrong. Let's say the person found an illness called hypothesitis, H. So the probability of the symptoms given hypothesitis is $P(E\backslash H) = 0.95$.

But Bayes' theorem can be used to find the probability that the person will have hypothesitis given the symptoms as

$$P(H\backslash E) = \frac{P(E\backslash H)P(H)}{P(E)}$$

The following information is still needed. The prior probability of hypothesitis $P(H)$, which is found to be a rare disease with probability $P(H) = 0.00001$. Also, the kind of symptoms does the person have, $P(E)$ (e.g., headache and cold) is $P(E) = 0.01$. So now $P(H\backslash E)$ is given by, using Equation 2.7, as

$$P(H\backslash E) = \frac{P(E\backslash H)P(H)}{P(E)} = \frac{0.95 \times 0.0001}{0.01} = 0.00095$$

Very small indeed!

2.1.4.1 Generalization of Bayes' Theorem

Often, for some partition $\{A_j\}$ of the sample space, the event space is given by or conceptualized in terms of $P(A_j)$ and $P(B\backslash A_j)$. It is then useful to compute $P(B)$ using the total law of probability as

$$P(B) = \sum_j P(B\backslash A_j)P(A_j)$$

which then gives a *generalized Bayes' theorem* as

$$P(A_i \backslash B) = \frac{P(B\backslash A_i)P(A_i)}{\sum_j P(B\backslash A_j)P(A_j)} \tag{2.8}$$

Example 2.7

The output from a factory is produced on three machines: A_1, A_2, and A_3. The three machines account for 20%, 30%, and 50% of the output, respectively. The fraction of defective items produced is 5% for A_1, 3% for A_2, and 1% for A_3. If an item is chosen at random and the output is found to be defective, what is the probability it was produced by machine A_3?

Solution: Let A_i denote the event that a randomly chosen item was made by the ith machine ($i = 1, 2, 3$). Let B denote the event that a randomly chosen item is defective. So we know $P(A_1) = 0.2$, $P(A_2) = 0.3$, and $P(A_3) = 0.5$. Also, $P(B\backslash A_1) = 0.05$, $P(B\backslash A_2) = 0.03$, and $P(B\backslash A_3) = 0.01$.

So using Bayes' theorem

$$P(A_3 \backslash B) = \frac{P(B \backslash A_3) P(A_3)}{\sum_{j=1}^{3} P(B \backslash A_j) P(A_j)}$$

$$P(A_3 \backslash B) = \frac{0.01 \times 0.5}{0.05 \times 0.2 + 0.03 \times 0.3 + 0.01 \times 0.5} = \frac{5}{24} = 0.208$$

2.1.5 Tree Diagrams

An alternative approach to solving probability problems involving a series of events is with a *tree diagram*. Consider the case of telephone calls made to an engineering supply company. The probability that a call is answered is given as 0.7. A tree diagram showing the possible outcomes is given in Figure 2.2. Each branch shows the possible outcomes of each call and their probabilities. Here, A = call answered and A' = call not answered. The outcome of the three calls is found by reading along the branches leading to it, and the probability of this outcome is found by multiplying the individual probabilities along these branches.

Note: This is an important concept in assessing the risk of certain events occurring. See examples in applications section.

From the tree diagram the probabilities of different outcomes can be calculated. The probability of all three calls being answered AAA can be seen to be 0.343. The probability of exactly two calls being answered is the sum of the probabilities of the three outcomes: AAA', $AA'A$, and $A'AA = 0.147 + 0.147 + 0.147 = 0.441$.

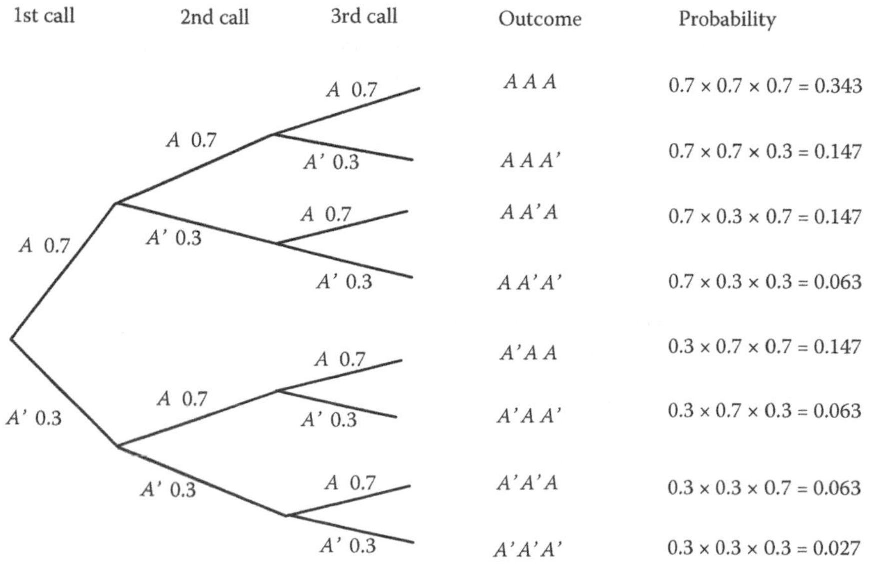

1st call	2nd call	3rd call	Outcome	Probability
		A 0.7	$A\,A\,A$	$0.7 \times 0.7 \times 0.7 = 0.343$
	A 0.7	A' 0.3	$A\,A\,A'$	$0.7 \times 0.7 \times 0.3 = 0.147$
		A 0.7	$A\,A'\,A$	$0.7 \times 0.3 \times 0.7 = 0.147$
A 0.7	A' 0.3	A' 0.3	$A\,A'\,A'$	$0.7 \times 0.3 \times 0.3 = 0.063$
		A 0.7	$A'\,A\,A$	$0.3 \times 0.7 \times 0.7 = 0.147$
	A 0.7	A' 0.3	$A'\,A\,A'$	$0.3 \times 0.7 \times 0.3 = 0.063$
A' 0.3	A' 0.3	A 0.7	$A'\,A'\,A$	$0.3 \times 0.3 \times 0.7 = 0.063$
		A' 0.3	$A'\,A'\,A'$	$0.3 \times 0.3 \times 0.3 = 0.027$

Figure 2.2 Tree diagram showing possible outcomes and probabilities.

2.2   Discrete Random Variables

2.2.1   Discrete Probability Distribution

Example 2.8

Consider a board game where a turn consists of throwing a die and then moving a number of squares equal to the score on the die. "The number of squares moved in a turn" is a variable because it can take different values, namely, 1, 2, 3, 4, 5, and 6.

However, the value taken at any one turn cannot be predicted but depends on chance. For these reasons "the number of squares moved in a turn" is called a *random variable*. Although the result of the next throw of the die cannot be predicted, it is known that, if the die is fair, the probability of getting each value is $\frac{1}{6}$.

Note: A random variable is a quantity whose value depends on chance.

A convenient way of expressing this information is to let X stand for the number of squares moved in a turn. Then, for example, $P(X = 3) = \frac{1}{6}$ means the probability that X takes the value 3 is $\frac{1}{6}$. Generalizing, $P(X = x)$ means the probability that the variable X takes the value x.

Note: The capital letter stands for the variable itself and the small letter stands for the value the variable takes.

This notation can be used in Table 2.1 below to give the possible values for the number of squares moved and the probability of each value. This table is called the probability distribution of X.

Note: The probability distribution of a random variable is just a listing of the possible values of the variable and the corresponding probabilities.

Table 2.1 Probability Distribution of X, the Number of Squares Moved in a Turn for a Single Throw of a Die

x	1	2	3	4	5	6	Total
$P(X = x)$	$\frac{1}{6}$	$\frac{1}{6}$	$\frac{1}{6}$	$\frac{1}{6}$	$\frac{1}{6}$	$\frac{1}{6}$	1

Example 2.9

A bag contains two red and three blue marbles. Two marbles are selected at random without replacement and the number, X, of blue marbles is counted. Find the probability distribution of X.

The tree diagram illustrating this situation is shown in Figure 2.3, where R_1 denotes the event that the first marble is red and R_2 the event that the second marble is red. Similarly, B_1 and B_2 stand for the events that the first and second marbles, respectively, are blue. X can take the values 0, 1, and 2.

$$P(X = 0) = P(R_1 \text{ and } R_2) = P(R_1) \times P\left(R_2\middle|R_1\right) = \frac{2}{5} \times \frac{1}{4} = \frac{2}{10} = \frac{1}{10}$$

$$\begin{aligned} P(X = 1) &= P(B_1 \text{ and } R_2) + P(R_1 \text{ and } B_2) \\ &= P(B_1) \times P\left(R_2\middle|B_1\right) + P(R_1) \times P\left(B_2\middle|B_1\right) \\ &= \frac{3}{5} \times \frac{2}{4} + \frac{2}{5} \times \frac{3}{4} = \frac{12}{20} = \frac{3}{5} \end{aligned}$$

$$P(X = 2) = P(B_1 \text{ and } B_2) = P(B_1) \times P\left(B_2\middle|B_1\right) = \frac{3}{5} \times \frac{2}{4} = \frac{6}{20} = \frac{3}{10}$$

The probability distribution of X is shown in Table 2.2.

Note: For any random variable, X, the sum of the probabilities is 1 and this is given as

$$\sum P(X = x) = 1 \qquad (2.9)$$

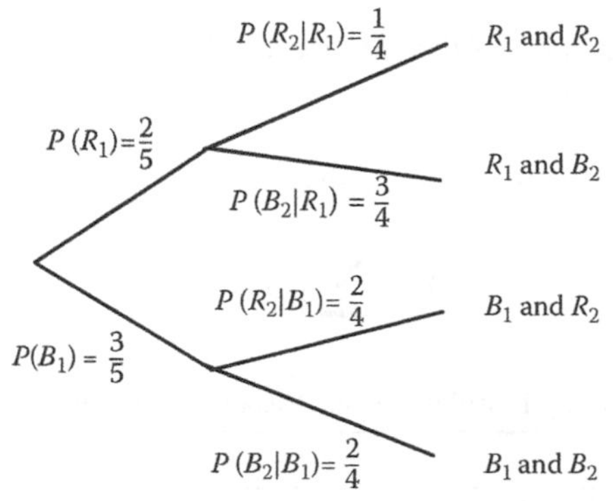

Figure 2.3 Tree diagram showing possible outcomes and probabilities.

Table 2.2 Probability Distribution of the Number of Blue Marbles X

x	0	1	2	Total
$P(X = x)$	$\frac{1}{10}$	$\frac{6}{10}$	$\frac{3}{10}$	1

2.2.2   Expectation Values

The expected value or the mean of a probability distribution is denoted by μ. The new symbol is used in order to distinguish the mean of a probability distribution from $\bar{x}$, the mean of a data set. μ is often called the **expectation** or *expected value* of X and is denoted by $E(X)$.

The expectation of a random variable X is defined by

$$E(X) = \mu = \sum x_i p_i \qquad (2.10)$$

Example 2.10

Find the expected value of the variable X, which has the probability distribution shown in Table 2.3.

Solution:

$$E(X) = \sum x_i p_i = \left(1 \times \frac{1}{6}\right) + \left(2 \times \frac{1}{6}\right) + \left(3 \times \frac{1}{6}\right)$$
$$+ \left(4 \times \frac{1}{6}\right) + \left(5 \times \frac{1}{6}\right) + \left(6 \times \frac{1}{6}\right) = 3.5$$

Table 2.3 Probability Distribution and Associated Probabilities

x	1	2	3	4	5	6	Total
$P(X=x)$	$\frac{1}{6}$	$\frac{1}{6}$	$\frac{1}{6}$	$\frac{1}{6}$	$\frac{1}{6}$	$\frac{1}{6}$	1

Example 2.11

Find the expected value of the variable Y, which has the probability distribution shown in Table 2.4.

Solution: Again using the formula gives

$$E(Y) = \sum y_i p_i = 4\frac{1}{12}$$

Table 2.4 Probability Distribution and Associated Probabilities

y	1	2	3	4	5	7	8	9	10	11	12	Total
$P(Y=y)$	$\frac{1}{6}$	$\frac{1}{6}$	$\frac{1}{6}$	$\frac{1}{6}$	$\frac{1}{6}$	$\frac{1}{36}$	$\frac{1}{36}$	$\frac{1}{36}$	$\frac{1}{36}$	$\frac{1}{36}$	$\frac{1}{36}$	1

2.2.3   Variance and Standard Deviation

Just as the spread in a data set can be measured by the standard deviation or variance, so it is possible to define a corresponding measure of a random variable. The symbol used for the standard deviation of a random variable is σ (read as "sigma") and its square, σ^2, is the variance of a random variable denoted by $Var(X)$.

The variance of a random variable X is defined as

$$\sigma^2 = Var(X) = \sum (x_i - \mu^2)p_i = \sum x_i^2 p_i - \mu^2 \qquad (2.11)$$

The standard deviation of a random variable is σ, the square root of $Var(X)$. It is in practice simpler to calculate $Var(X)$ using $\sum x_i^2 p_i - \mu^2$.

Example 2.12

Calculate the standard deviation of the random variable X given in Example 2.10.

Solution: First calculate $\sum x_i^2 p_i$:

$$\sum x_i^2 p_i = \left(1^2 \times \frac{1}{6}\right) + \left(2^2 \times \frac{1}{6}\right) + \left(3^2 \times \frac{1}{6}\right) + \left(4^2 \times \frac{1}{6}\right)$$
$$+ \left(5^2 \times \frac{1}{6}\right) + \left(6^2 \times \frac{1}{6}\right) = \frac{91}{6} = 15\frac{1}{6}$$

From the previous Example 2.10, $\mu = E(X) = 3.5$. Using definition of variance

$$Var(X) = \sum x_i^2 p_i - \mu^2 = 15\frac{1}{6} - 3.5^2 = \frac{35}{12} = 2.92$$

Then calculate the standard deviation, which is just the square root of the variance:

$$\sigma = \sqrt{\frac{35}{12}} = 1.71 \text{ (correct to three significant figures)}$$

2.3 Continuous Random Variables

Note: In this section, finding the integral of functions is required and so the topic of integration is important. If one is not familiar with integrating functions, then it is advised to study Chapter 6 first, then return to this section afterward.

It has already been shown that for a discrete random variable X, it is possible to allocate probabilities to each discrete value, x, that X can take. When considering a continuous random variable this is not the case.

For a continuous random variable X, probabilities are allocated to each of the range of values that the variable can take. This is done by defining a function $f(x)$ called the *probability density function* (pdf).

2.3.1 Probability Density Function (pdf)

The probability density function $f(x)$ allocates probabilities to each of the range of values that the continuous random variable can take and is defined such that for $f(x) \geq 0$, for all values of x, then

$$\int_{all\,x} f(x)\,dx = 1 \tag{2.12}$$

Then, the probability that the random variable X takes a value in the range $a \leq x \leq b$ is given by the integral

$$P(a \leq X \leq b) = \int_{x=a}^{x=b} f(x)\,dx \tag{2.13}$$

Example 2.13

A continuous random variable X has the pdf defined by

$$f(x) = \begin{cases} \dfrac{3}{4}(1 - x^2) & -1 \leq x \leq 1 \\ 0 & \text{otherwise} \end{cases}$$

Calculate $P(0.2 \leq X \leq 0.5)$.

Solution:

$$P(0.2 \leq X \leq 0.5) = \int_{0.2}^{0.5} \frac{3}{4}(1 - x^2)\,dx$$

$$\frac{3}{4}\left[x - \frac{x^3}{3}\right]_{0.2}^{0.5} = \frac{3}{4}\left[\left(0.5 - \frac{0.5^3}{3}\right) - \left(0.2 - \frac{0.2^3}{3}\right)\right]$$

$$= 0.196 \text{ (correct to three significant digits)}$$

2.3.2 Cumulative Distribution Function (cdf)

The cumulative distribution function, $F(x)$, for a continuous random variable, X, having a pdf $f(x)$ is given by the formula

$$F(x) = P(X \leq x) = \int_{-\infty}^{x} f(x)\,dx \tag{2.14}$$

where $\dfrac{dF(x)}{dx} = f(x)$.

Note: If f(x) is defined only on the range of values $a \leq x \leq b$, then this becomes

$$F(x) = P(X \le x) = \int_a^x f(x)\,dx \qquad (2.15)$$

Example 2.14

The continuous random variable X has the following cumulative distribution function:

$$F(x) = \begin{cases} 0 & x \le 0 \\ \dfrac{x^3}{64} & 0 < x \le 4 \\ 1 & x > 4 \end{cases}$$

a. Find $P(X \le 3)$.
b. Sketch the graph of $f(x)$.

Solution:

a. $P(X \le 3) = F(3) = \dfrac{3^3}{64} = \dfrac{27}{64}$.

b. Using, $f(x) = \dfrac{dF(x)}{dx}$

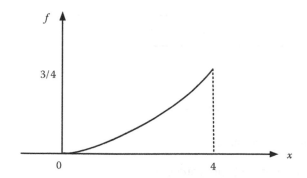

Figure 2.4 Graph of the probability density function $f(x)$.

The pdf is shown in Figure 2.4 and defined as

$$f(x) = \begin{cases} \dfrac{3x^2}{64} & 0 \le x \le 4 \\ 0 & \text{otherwise} \end{cases}$$

2.3.3   Expectation of a Continuous Random Variable

The expected value of a continuous random variable X having a pdf $f(x)$ is denoted by $E(X)$, where

$$E(X) = \int_{allx} x f(x) \, dx \qquad (2.16)$$

Example 2.15

A random variable T has the pdf given by

$$f(t) = \begin{cases} 0.02t & 0 < x < 10 \\ 0 & \text{otherwise} \end{cases}$$

Find $E(T)$.

Solution: Using the formula given by Equation 2.16 gives

$$E(T) = \int_0^{10} t f(t) \, dt$$

$$= \int_0^{10} 0.02t^2 \, dt$$

$$= \left[\frac{0.02t^3}{3} \right]_0^{10} = 6\frac{2}{3}$$

Note: In general, the following results can be shown to be true.

$$E(aX) = aE(X) \qquad (2.17)$$

$$E(aX + b) = aE(X) + b \qquad (2.18)$$

where a and b are constants.

2.3.4   Variance and Standard Deviation of a Continuous Random Variable

The variance of a continuous random variable X is defined as

$$Var(X) = E(X^2) - \left[E(X) \right]^2 \qquad (2.19)$$

Usually the variance is $Var(X) = \sigma^2$.

The standard deviation of a continuous random variable X is denoted by σ and is given by

$$\sigma = \sqrt{Var(X)} \qquad (2.20)$$

Example 2.16

A random variable T has the pdf given by

$$f(t) = \begin{cases} 0.02t & 0 < x < 10 \\ 0 & \text{otherwise} \end{cases}$$

Find the variance $Var(T)$ and the standard deviation σ.

From the previous Example 2.15 it was found that $E(T) = 6\frac{2}{3}$.

Solution:

$$Var(T) = E(T^2) - \left[E(T) \right]^2$$

So

$$E(T^2) = \int_0^{10} t^2 f(t)\, dt$$

$$= \int_0^{10} 0.02t^3 \, dt$$

$$= \left[\frac{0.02t^4}{4} \right]_0^{10} = 50$$

Therefore,

$$Var(T) = 50 - \left[6\frac{2}{3} \right]^2 = 5\frac{5}{9}$$

And the standard deviation is

$$\sigma = \sqrt{5\frac{5}{9}} = 2.36$$

Note: In general, the following results can be shown to be true for the variance of a continuous random variable X:

$$Var(a) = 0 \qquad\qquad (2.21)$$

$$Var(aX) = a^2 \; Var(X) \tag{2.22}$$

$$Var(aX + b) = a^2 \; Var(X) \tag{2.23}$$

where a and b are constants.

2.4   Applications

Example 2.17: Event Tree Analysis (ETA) Showing Probabilities of Outcomes

Failure of a complicated engineering system can lead to different damage scenarios. The consequence of a particular failure event may depend on a sequence of events following the failure. The means for systematic identification of the possible event sequences is the so-called *event tree*. This is a visual representation, indicating all events that can lead to different scenarios. In the following example, first identify the events.

Consider the initiation event A, fire ignition reported to a fire squad. After the squad has been alerted and done its duty at the place of accident, a form is completed where a lot of information about the fire can be found: type of alarm, type of building, number of staff involved, and much more. Here the focus is on the condition of the fire at the arrival of the fire brigade. This is described as

E_1: Smoke production without flames

and the complement as

E_1^c: A fire with flames (not merely smoke production)

The place where the fire was extinguished is described by the event

E_2: Fire was extinguished in the item where it started

and the complement as

E_2^c: Fire was extinguished outside the item

Figure 2.5 shows the events and corresponding number of cases.

Consider the case, where there was a fire with flames at the arrival and that the fire was extinguished outside the place where it started. The probabilities can be calculated from the event tree or using the conditional probability formula Equation 2.6.

From the event tree this is

$$E_1^c \; E_2^c = \frac{65}{100} \times \frac{30}{65} = \frac{30}{100} = 0.3$$

or using Equation 2.6 gives

$$P\left(E_1^c \text{ and } E_2^c\right) = P\left(E_1^c\right)P\left(E_2^c\big|E_1^c\right) = \frac{65}{100}\times\frac{30}{65} = \frac{30}{100} = 0.3$$

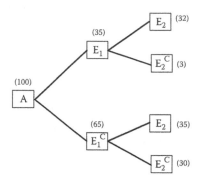

Figure 2.5 The event tree with the numbers within the parentheses indicating the number of cases observed after 100 fire ignitions.

Example 2.18: Fault Tree Analysis (FTA) to Calculate Risk Assessment

The fault tree analysis method can be used as an essential element for risk assessment studies and accident investigations. The method starts with a schematic diagram representing the system components, which have an associated probability of failure attached to them and uses a bottom-up method to analyze the system failure (i.e., the top event). The diagram consists of the system components combined with the logic operators the "And" and "Or" gates. The "And" gate implies that for two components the system will only fail if the first component and the second fails. For an "Or" gate, the system will fail if either the first or second component fails.

As an example, when tackling fire incidents firefighters need a constant supply of water to deal with the fire. The top event here can be that there is no water output for dealing with the incident. The water supply comes via an electrical pump system and a diesel pump system. The electrical pump system can fail due to the pump or the electrical supply, and similarly, the diesel pump system can fail due to the pump or diesel supply. Each system component has an associated probability of failure and working. From the bottom up a total probability failure can be calculated for the top event using the rules of probabilities for "And" and "Or" systems. The preceding system can be represented by a qualitative and quantitative fault tree as shown Figure 2.6.

Starting with the bottom probabilities and using the rules for "And" and "Or" the higher probabilities can be calculated so that the top event failure rate can be determined. If the fire company is not happy with the failure rate for the top event of 0.01206 or 1/83 failures, then further action can be taken to increase the overall reliability of the system.

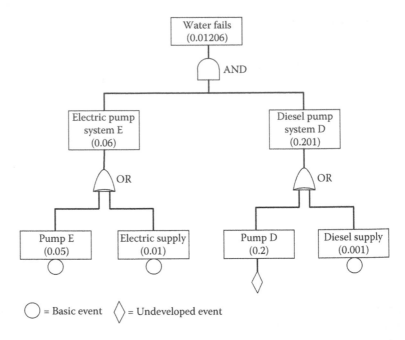

Figure 2.6 Fault tree diagram showing the probabilities of failure in parentheses.

Here, the diesel pump D is the most unreliable component and so making this more reliable (i.e., reducing its failure rate to 0.1) would mean the top event failure rate now becomes 0.00606 or 1/165 failures, which is a good improvement. Even further analysis could be carried out by the company to reduce the risk of failure by introducing a second (i.e., a spare) diesel pump and seeing what the effect of this would have on the overall failure rate. It turns out that introducing "redundancy" is a more effective method for increasing the overall reliability of a system (see Problem 2.8).

**Example 2.19: Application of Bayes' Theorem
for Conditional Probabilities**

Three companies X, Y, and Z supply sprinkler systems to a university. The percentage of sprinklers supplied and the probability of these being defective is shown in Table 2.5.

Given that a sprinkler is found to be defective, what is the probability of it being supplied by company Y?

Table 2.5 Percentage of Sprinkler's Supplied and Probability of Being Defective

Company	Percentage of Sprinklers Supplied	Probability of Being Defective
X	60%	0.01
Y	30%	0.02
Z	10%	0.03

Using Bayes' theorem, that is, Equation 2.8, gives

$$P(Y\backslash D) = \frac{P(Y)P(D\backslash Y)}{P(X)P(D\backslash X) + P(Y)P(D\backslash Y) + P(Z)P(D\backslash Z)}$$

$$P(Y\backslash D) = \frac{0.3 \times 0.02}{0.6 \times 0.01 + 0.3 \times 0.02 + 0.1 \times 0.03} = \frac{0.006}{0.015} = 0.4$$

that is, there is a 40% chance that the defective sprinkler was supplied by company Y.

Problems

2.1 A card is picked at random from a shuffled pack. What is the probability of getting

 a. A diamond

 b. An ace or a king

 c. An ace or a red card

2.2 A group of firefighters is called to a fire incident. Seven are full-time, five are part-time, and two are reserves. If a firefighter is chosen at random to drive the fire engine, what is the probability it will be a part-time firefighter?

2.3 Two used fire engines were bought by a fire station. The probability that fire engine A is working in a year's time is 0.9 and the probability that the second fire engine B is working in a year's time is 0.7. Find the probability that

 a. Both fire engines are working in a year's time.

 b. At least one of the fire engines is still working in a year's time.

2.4 A fair coin is tossed three times. Find the probability that the number of tails is 0, 1, 2, or 3.

2.5 A high-rise building decides, retrospectively, to install smoke alarms in all its apartments. It buys the smoke alarms from two companies: company A supplies 65% and company B supplies 35% of the alarms. The probability that a smoke alarm is defective from company A is 0.01 and from company B is 0.03. After routine testing, it was found that a smoke alarm was defective. What is the probability it was supplied by company B?

2.6 Calculate the expected value and the standard deviation of the variable X, which has the probability distribution given in Table 2.6.

Table 2.6 Probability Distribution
and Associated Probabilities

x	1	2	3	Total
$P(X = x)$	$\dfrac{1}{6}$	$\dfrac{3}{6}$	$\dfrac{2}{6}$	1

2.7 Given that a system has a failure probability density function $f(t)$, where time is measured in years as follows,

$$
f(t)\begin{cases} 0 & t < 0 \\[2mm] \dfrac{1}{8}t & 0 \le t \le 4 \\[2mm] 0 & t > 4 \end{cases}
$$

Calculate the mean time to failure (MTTF) and the standard deviation for the system.

2.8 Consider the system given in Example 2.18. It was found that reducing the diesel pump failure rate by half to 0.1 reduced the overall failure rate from 1/83 to 1/165. To further improve this system, the company decides to install a second pump (i.e., a spare) with the same failure rate of 0.2 rather than improving the reliability of the first pump. Calculate the new system failure rate for the top event "water fails" and comment on your findings.

③ Vectors and Geometrical Applications

3.1 Introduction

In engineering, there are different mathematical quantities that are used to describe physical phenomena. These quantities can be divided into two categories: scalar or vector. These two quantities are defined as follows:

Scalar quantity—Something that has only size/magnitude. There are many examples of these, such as time, temperature, volume, mass, and speed (usually a number and units).

Vector quantity—Something that has both size/magnitude and direction. Examples of these are force, velocity, and acceleration.

Note: A vector is usually distinguished from a scalar by a line either on top of or below the symbol, that is, as $\bar{a}$ or $\underline{a}$. Both notations are used in this book.

To represent vectors in two- or three-dimensional space, coordinate axes are used with unit vectors defined along each axis as follows:

A unit vector is any vector that has size (or magnitude) equal to one. In the x, y, z direction of space these unit vectors are defined as follows: $\underline{i}, \underline{j}$ are unit vectors in the x and y directions in 2-D, respectively, and so vector $\underline{a} = \overline{3}\underline{i} + 4\underline{j}$ is given by three units along the x-axis and four units along the y-axis and is shown in Figure 3.1. Similarly, $\underline{i}, \underline{j}$, and $\underline{k}$ are unit vectors in the x, y, and z directions in 3-D, respectively, and so a vector $\underline{b} = 2\underline{i} + 3\underline{j} + 7\underline{k}$ is shown in Figure 3.2.

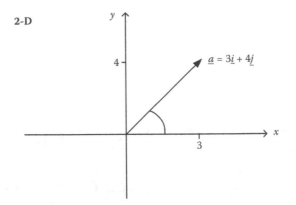

Figure 3.1 Vector represented in 2-D space.

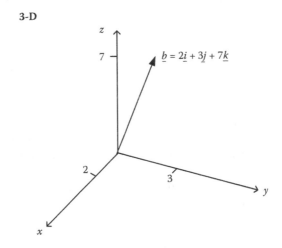

Figure 3.2 Vector represented in 3-D space.

3.1.1   Magnitude and Unit Vectors

Giving a vector $\underline{r}$, the size or magnitude of the vector $\underline{r}$ is denoted by $\left|\underline{r}\right|$ and shown in Figure 3.3. The magnitude is calculated geometrically using the Pythagorean theorem in 2-D and 3-D space. Generally, for the vector $\underline{r} = a\underline{i} + b\underline{j}$, the magnitude $\left|\underline{r}\right|$ is given by $\left|\underline{r}\right| = \sqrt{a^2 + b^2}$, that is, the square root of the sum of the components squared.

In 2-D, the angle of the vector , that is, the argument, is given by $\theta = \tan^{-1}\left(\dfrac{b}{a}\right)$, but more generally if the vector is in a different quadrant, then trigonometry is used to calculate θ.

Given a general vector in 3-D space, $\underline{r} = x\underline{i} + y\underline{i} + z\underline{k}$, as shown in Figure 3.4. Generally, for the vector $\underline{r} = x\underline{i} + y\underline{i} + z\underline{k}$ the magnitude $\left|\underline{r}\right|$ is given as $\left|\underline{r}\right| = \sqrt{x^2 + y^2 + z^2}$, again the square root of the sum of the components squared.

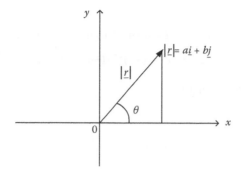

Figure 3.3 General vector with its magnitude in 2-D space.

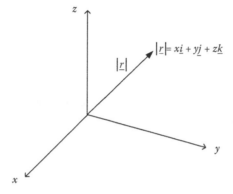

Figure 3.4 General vector with its magnitude in 3-D space.

Example 3.1

In 2-D space:

If $\underline{r} = 3\underline{i} + 4\underline{j}$, then the magnitude of $\underline{r}$ is given as $|\underline{r}| = \sqrt{3^2 + 4^2}$, $|\underline{r}| = \sqrt{25}$, so $|\underline{r}| = 5$.

Example 3.2

In 3-D space:

Given the vector $\underline{r} = \underline{i} + \underline{j} - 4\underline{k}$, find the magnitude $|\underline{r}|$.

$$|\underline{r}| = \sqrt{(1)^2 + (1)^2 + (-4)^2}, |\underline{r}| = \sqrt{18}, \text{ so } |\underline{r}| = 4.24.$$

3.1.1.1 Unit Vectors

Given a vector $\underline{r}$, a unit vector in the direction of $\underline{r}$ is denoted by $\hat{\underline{r}}$, where $\hat{\underline{r}} = \dfrac{\underline{r}}{|\underline{r}|}$.

This means that to find a unit vector, take the vector and divide its components by the magnitude of that vector.

Example 3.3

For the vector given by $r = 2i - 4j + 3k$ to find the unit vector in the direction of r, using $\hat{r} = \dfrac{r}{|r|}$ gives $|r| = \sqrt{(2)^2 + (-4)^2 + (3)^2}$, $|r| = \sqrt{4 + 16 + 9}$ and so $|r| = \sqrt{29}$. This then gives

$$\hat{r} = \frac{1}{\sqrt{29}} \cdot (2i - 4j + 3k)$$

As will be seen in later chapters vectors have very important applications in engineering in areas such as *vector fields*. An example of a basic vector field is shown next.

Example 3.4

What does the vector field defined in 2-D space given by $F(x, y) = x\,i$ look like? To see this, take different points in space and see what the vector field becomes.

So for the point (1,0), $F(1,0) = 1\,i$ and for any point (1, y), $F(1, y) = 1\,i$. Similarly, for the point (2, y), $F(2, y) = 2\,i$. This vector field is shown in Figure 3.5.

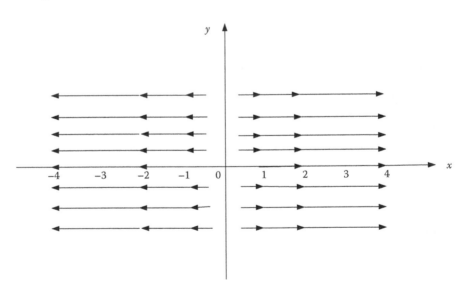

Figure 3.5 Vector field defined by $F(x, y) = x\,i$.

3.1.2   Addition and Subtraction of Vectors

Addition of vectors is carried out with simple adding of the corresponding i, j, and k components. Figure 3.6 shows how to add vectors pictorially.

To calculate the vector $c = a + b$, just add the corresponding components as follows: Given $a = 3i + 2j - 5k$ and $b = i - 3j + 7k$, then $c = a + b = 4i - j + 2k$.

Subtraction of vectors is carried out with subtracting of corresponding i, j, and k components. Figure 3.7 shows how to subtract vectors pictorially.

Figure 3.6 Addition of two vectors $\underline{a}$ and $\underline{b}$.

Figure 3.7 Subtraction of two vectors $\underline{a}$ and $\underline{b}$.

To calculate the vector $\underline{d} = \underline{a} - \underline{b}$, just subtract the corresponding components. Given $\underline{a} = 3\underline{i} + 2\underline{j} - 5\underline{k}$ and $\underline{b} = \underline{i} - 3\underline{j} + 7\underline{k}$, then $\underline{d} = \underline{a} - \underline{b} = 2\underline{i} + 5\underline{j} - 12\underline{k}$.

3.1.3   Scalar and Vector Products

First, consider what is meant by multiplication of a vector by a scalar k. In the following equation, $\underline{a} = k\,\underline{b}$, this means that the vector $\underline{a}$ is k times as long as the vector $\underline{b}$ and in the same direction. If k happens to be negative, then they are in the opposite direction.

When considering the product of two vectors, there is some freedom to what is meant by this. Generally, there are two different products that are defined: one called the *scalar product* since the result is a scalar quantity and the other is the *vector product* as this produces a vector after the multiplication.

3.1.3.1   Scalar Product (Dot Product)

Consider two vectors $\underline{a}$ and $\underline{b}$ and an angle θ between them as shown in Figure 3.8. The definition of the scalar product is given by

$$\underline{a} \cdot \underline{b} = |\underline{a}||\underline{b}|\cos\theta = \underline{b} \cdot \underline{a} \qquad (3.1)$$

This can be thought of as the length of $\underline{a}$ multiplied by the component of $\underline{b}$ along $\underline{a}$ or vice versa.

Note: The scalar product is always written with a dot between the two vectors being multiplied.

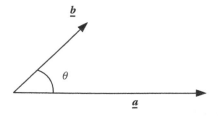

Figure 3.8 Two vectors with an angle between them in space.

The formula given by Equation 3.1 is useful for the following situations: (1) to find the angle between two vectors, and (2) to prove if two vectors are perpendicular to each other.

Consider the arbitrary vectors $\underline{a} = a_1\underline{i} + a_2\underline{j} + a_3\underline{k}$ and $\underline{b} = b_1\underline{i} + b_2\underline{j} + b_3\underline{k}$. To calculate the $\underline{a} \cdot \underline{b}$ term, this is given as

$$\underline{a} \cdot \underline{b} = a_1b_1 + a_2b_2 + a_3b_3 \tag{3.2}$$

This is because the terms $\underline{i} \cdot \underline{i} = \underline{j} \cdot \underline{j} = \underline{k} \cdot \underline{k} = 1$ and $\underline{i} \cdot \underline{j} = \underline{j} \cdot \underline{k} = \underline{i} \cdot \underline{k} = 0$.

Example 3.5

Given $\underline{a} = 3\underline{i} + 2\underline{j} + \underline{k}$ and $\underline{b} = \underline{i} + 5\underline{j} - 3\underline{k}$, find the angle between the vectors. Using Equation 3.2 first gives

$$\underline{a} \cdot \underline{b} = (3)(1) + (2)(5) + (1)(-3)$$
$$= 3 + 10 - 3 = 10$$

Now, to find the angle θ between the vectors using Equation 3.1 gives

$$\underline{a} \cdot \underline{b} = |\underline{a}||\underline{b}|\cos\theta$$

$$10 = \sqrt{3^2 + 2^2 + 1^2} \cdot \sqrt{1^2 + 5^2 + (-3)^2} \cdot \cos\theta$$

$$\cos\theta = \frac{\underline{a} \cdot \underline{b}}{|\underline{a}| \cdot |\underline{b}|}, \quad \cos\theta = \frac{10}{\sqrt{14} \cdot \sqrt{35}} = \frac{\sqrt{10}}{7} = 0.452$$

$$\theta = \cos^{-1}\left(\frac{\sqrt{10}}{7}\right) = 63.14°$$

Note: If $\underline{a} \cdot \underline{b} = 0$, then the two vectors are perpendicular, since $|\underline{a}|$ and $|\underline{b}| \neq 0$, so $\cos\theta = 0$, which implies that the angle $\theta = 90°$.

3.1.3.2   Vector Product (Cross Product)

The definition of the vector product is given as

$$\underline{a} \times \underline{b} = |\underline{a}||\underline{b}|\sin\theta\,\hat{n} \tag{3.3}$$

where $\hat{n}$ is a unit normal vector to the plane containing both $\underline{a}$ and $\underline{b}$. There are two directions perpendicular to any plane, here the direction is such that $\underline{a}$, $\underline{b}$, and $\hat{n}$ form a right-handed set of axes.

Note: The vector product can have a "×" or a "∧" notation between the vectors, Figure 3.9 shows the arrangement of the vectors.

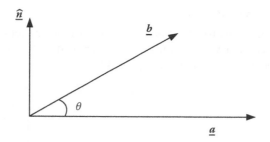

Figure 3.9 The vector product of $\underline{a}$ and $\underline{b}$ lies along the z axis.

Equation 3.3 for the vector product is useful for producing a vector that is perpendicular to both the vectors $\underline{a}$ and $\underline{b}$.

3.1.3.3 How to Calculate $\underline{a} \times \underline{b}$

If two vectors are given as $\underline{a} = 3\underline{i} + 2\underline{j} + \underline{k}$ and $\underline{b} = \underline{i} + 2\underline{j} - 3\underline{k}$, then the vector product is given as $\underline{a} \times \underline{b} = (3\underline{i} + 2\underline{j} + \underline{k}) \times (\underline{i} + 2\underline{j} - 3\underline{k})$. Now this can be slightly complicated to work out by expanding of the brackets and using $\underline{i} \times \underline{i} = \underline{j} \times \underline{j} = \underline{k} \times \underline{k} = 0$, while the cross products follow the right-handed axes rules. This process can be seen using the cyclic diagram as shown in Figure 3.10.

However, the vector product is more easily calculated using a 3×3 determinant. Since what is required is the vector $\underline{c} = \underline{a} \times \underline{b}$, in the determinant the first row is just the $\underline{i}, \underline{j}$, and $\underline{k}$ components. Then on the second row will be the components of the vector $\underline{a}$ and finally on the third row the components of the vector $\underline{b}$ as follows:

$$\underline{c} = \underline{a} \times \underline{b} = \begin{vmatrix} \underline{i} & \underline{j} & \underline{k} \\ 3 & 2 & 1 \\ 1 & 2 & -3 \end{vmatrix} = +\underline{i} \begin{vmatrix} 2 & 1 \\ 2 & -3 \end{vmatrix} - \underline{j} \begin{vmatrix} 3 & 1 \\ 1 & -3 \end{vmatrix} + \underline{k} \begin{vmatrix} 3 & 2 \\ 1 & 2 \end{vmatrix}$$

$$\underline{c} = -8\underline{i} + 10\underline{j} + 4\underline{k}$$

Now, if this is a vector perpendicular to both $\underline{a}$ and $\underline{b}$, then this can be checked using the scalar product of $\underline{c}$ with $\underline{a}$ and $\underline{b}$. The result should be zero.

Check: $\underline{a} \cdot \underline{c} = 0$ $-33 + 20 + 13 = 0$

Also, $\underline{b} \cdot \underline{c} = 0$ $-8 + 20 - 12 = 0$

This proves that both vectors are perpendicular to this vector $\underline{c} = \underline{a} \times \underline{b}$.

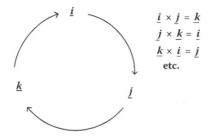

Figure 3.10 Cyclic rotation showing vector products for unit vectors $\underline{i}, \underline{j}$, and, $\underline{k}$.

3.1.4   Projection of Vectors

When talking about the projection of a vector onto another vector, this generally means the projection of vector $\underline{a}$ on to the line created by the vector $\underline{b}$. The following cases could arise as shown in Figure 3.11.

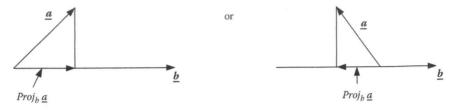

Figure 3.11 Projection of vector $\underline{a}$ onto $\underline{b}$.

This idea is similar to a sundial casting a shadow on a tabletop.

How do you find the projection of $\underline{a}$ onto $\underline{b}$? Consider the situation in Figure 3.12.

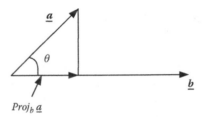

Figure 3.12 Calculating the projection of the vector $\underline{a}$ onto $\underline{b}$.

First find the length of the projection of $\underline{a}$ on to $\underline{b}$. This is given by using trigonometry. The length of projection onto $\underline{b}$ is $= |\underline{a}| \cos\theta$. This is in terms of the angle θ. But in terms of the vectors $\underline{a}$ and $\underline{b}$, make use of the dot product as follows:

$$\underline{a} \cdot \underline{b} = |\underline{a}||\underline{b}| \cos\theta$$

gives

$$\cos\theta = \frac{\underline{a} \cdot \underline{b}}{|\underline{a}||\underline{b}|}$$

Therefore, replacing for cos θ, the length of the projection of $\underline{a}$ on to $\underline{b}$ is now

$$|\underline{a}| \frac{\underline{a} \cdot \underline{b}}{|\underline{a}||\underline{b}|} = \frac{\underline{a} \cdot \underline{b}}{|\underline{b}|}$$

So the length of the projection vector is $\dfrac{\underline{a} \cdot \underline{b}}{|\underline{b}|}$, but its direction is that of the vector $\underline{b}$. Therefore, multiplying the length by a unit vector $\hat{\underline{b}}$ will give the vector projection of $\underline{a}$ onto $\underline{b}$ as follows:

$$Proj_{\underline{b}} \, \underline{a} = \frac{\underline{a} \cdot \underline{b}}{|\underline{b}|} \hat{\underline{b}} \qquad\qquad (3.4)$$

Now, $\dfrac{b}{|b|} = \hat{b}$ can be written out again in a neater form as

$$Proj_b \, \underline{a} = \left(\underline{a}.\hat{\underline{b}}\right)\hat{\underline{b}} \qquad (3.5)$$

Note: This can be useful when imagining a fire hose spraying foam from the ground toward a fire on an oil tank. This gives a way to combine the horizontal wind speed with the form jet in that direction.

Example 3.6

Find the distance from the point (5, 5) to the line $y = 2x$ shown in Figure 3.13. In this problem, the use of projections can help to reach a solution.

Solution: To find the perpendicular distance d of the point (5, 5) to the line, first project the vector (5, 5) onto the line. A vector that describes the line $y = 2x$ is given as the vector $\langle 1,2\rangle$. You need to find

$$Proj_{\langle 1,2\rangle}\langle 5,5\rangle = \left(\langle 5,5\rangle.\frac{\langle 1,2\rangle}{\sqrt{5}}\right)\left(\frac{\langle 1,2\rangle}{\sqrt{5}}\right)$$

$$= \frac{15}{5}\langle 1,2\rangle = 3\langle 1,2\rangle = \langle 3,6\rangle$$

So the projection of $\langle 5,5\rangle$ onto the line is at the point P (3, 6). This is the closest point to (5, 5) that exists on the line $y = 2x$.

So the distance required is the distance between two points (5, 5) and (3, 6), which is just given by the distance between two points formula, $d = \sqrt{(x_1 - x_2)^2 + (y_1 - y_2)^2}$, that is, $d = \sqrt{(5-3)^2 + (5-6)^2} = \sqrt{5}$.

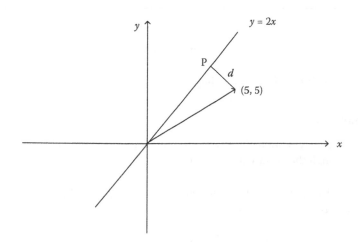

Figure 3.13 Perpendicular distance from a point to the line.

3.2   Vector Geometry

3.2.1   Vector Equation of a Line

To find the vector equation of the straight line shown in Figure 3.14, suppose P is an arbitrary point on a straight line and A and B are given points on the same line and O is the origin. P has a position vector $\underline{r}$, the line is parallel (in the direction) of $\underline{AB}$, and A and B are both points on the line.

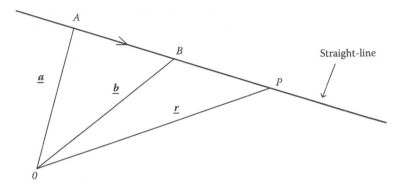

Figure 3.14 Vector equation of a line.

Now using vector addition gives the relationship

$$\underline{OP} = \underline{OA} + \underline{AP}$$

where $\underline{OP} = \underline{r}$ (variable), $\underline{OA} = \underline{a}$ (fixed), and $\underline{AP} = t\underline{AB}$ (t is a real parameter). Therefore, the *vector equation* of a straight line can be written as

$$\underline{r} = \underline{a} + t\underline{AB} \tag{3.6}$$

Note: The vector equation of a straight line through two fixed points with position vectors $\underline{a}$ and $\underline{b}$ is given by

---

$$\underline{r} = \underline{a} + t(\underline{b} - \underline{a}) \tag{3.7}$$

(Since $\underline{AB} = \underline{OA} + \underline{OB}$ from the diagram.)

Example 3.7

Find the vector equation of the straight line parallel to the vector $\underline{i} + 2\underline{j} - 5\underline{k}$ going through the point with position vector $2\underline{i} - 3\underline{j} + \underline{k}$.

Solution: Using the vector equation of a line is given by Equation 3.6 $\underline{r} = \underline{a} + t\underline{AB}$:

$$\underline{r} = 2\underline{i} - 3\underline{j} + \underline{k} + t(\underline{i} + 2\underline{j} - 5\underline{k})$$

3.2.1.1 Intersection of Lines

In 2-D space two lines intersect or are parallel. In 3-D space two lines can intersect or they may be either parallel or skew (i.e., missing each other).

Given two lines in space $r_1 = a_1 + t b_1$ and $r_2 = a_2 + s b_2$, then for them to intersect means $r_1 = r_2$ (i.e., $a_1 + t b_1 = a_2 + s b_2$), where the unique values of t and s can be found.

Note: If unique values of t and s cannot be found, then the lines do not intersect and are said to be skew.

Example 3.8

Find the point of intersection of the lines:

$$r_1 = 2i + 3j + t(-i + 2j) \quad \text{and} \quad r_2 = -i + j + s(3i - 2j)$$

Solution: Lines intersect when $r_1 = r_2$

$$2i + 3j + t(-i + 2j) = -i + j + s(3i - 2j)$$

$$(2 - t)i + (3 + 2t)j = (-1 + 3s)i + (1 - 2s)j$$

Equating the coefficients of the i and j components gives

$$2 - t = -1 + 3s \quad (\text{i.e. } 3s + t = 3)$$

and

$$3 + 2t = 1 - 2s \quad (\text{i.e. } t + s = -1)$$

Solving these simultaneously gives $t = -3$ and $s = 2$.

Therefore, the lines do meet and the point of intersection is given by using either the $t = -3$ or $s = 2$ back into the equations of the lines r_1 or r_2, respectively. Using $t = -3$ into $r_1 = 2i + 3j + t(-i + 2j)$ gives $2i + 3j - 3(-i + 2j) = 5i - 3j$ as the *point of intersection* of the two lines.

Note: If the lines are in 3-D space, follow the same method except now there will be three simultaneous equations involving s and t. Solve for s and t using any two of the three equations and test to see if these fit into the third equation. If they fit into the third equation, then the lines do intersect. If they do not fit the third equation, then the lines are said to be skew.

Example 3.9

Two lines r_1 and r_2 are given by the equations:

$$r_1 = 3\underline{i} + \underline{j} - 4\underline{k} + s(-2\underline{i} - 3\underline{j} + 6\underline{k})$$

$$r_2 = 2\underline{i} - 3\underline{j} + \underline{k} + t(\underline{i} - \underline{j} - \underline{k})$$

where s and t are parameter values.

Show that these lines intersect and find the position of the vector of P, the point of intersection.

Solution: For intersection of the lines $r_1 = r_2$, this gives

$$(3 - 2s)\underline{i} + (1 - 3s)\underline{j} + (-4 + 6s)\underline{k} = (2 + t)\underline{i} + (-3 - t)\underline{j} + (1 - t)\underline{k}$$

Equating the coefficients of the $\underline{i}, \underline{j}$, and $\underline{k}$ components gives

$$3 - 2s = 2 + s, \quad 1 - 3s = -3 - t, \quad -4 + 6s = 1 - t$$

Solving the first two equations gives $s = 1$ and $t = -1$, and these fit the third equation so therefore the lines do intersect.

The point of intersection is given by using $s = 1$ into equation for r_1 and the point becomes $P(\underline{i} - 2\underline{j} + 2\underline{k})$.

3.2.2   Vector Equation of Planes

A plane in space is completely specified by knowing one point in it, together with a vector that is perpendicular to the plane.

Example 3.10

Find the equation of the plane with normal vector $\underline{n} = \langle -1, 2, 3 \rangle$ and goes through the point Q (1, 3, 1) as shown in Figure 3.15.

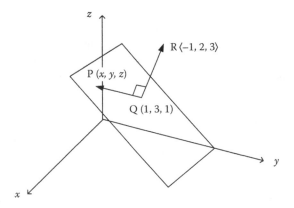

Figure 3.15 Equation of a plane, normal to a vector and passing through a point.

To find this plane, for a given point (x, y, z) the vector $\underline{PQ}$ and $\underline{QR}$ must be perpendicular to each other. So, using the properties of the scalar product, then the dot product of $\underline{QR}$ with $\underline{PQ}$ must be zero.

Therefore, $\langle -1, 2, 3 \rangle \cdot \langle x-1, y-3, z-1 \rangle = 0$

$$-(x-1) + 2(y-3) + 3(z-1) = 0$$

the equation of the plane is $-x + 2y + 3z = 8$.

3.2.2.1   Generalizing for Any Plane in Space

First, one needs a single point that the plane goes through and a normal vector $\underline{n}$ to the plane. Let the point in the plane be (x_0, y_0, z_0) and the normal vector be $\underline{n} = \langle a, b, c \rangle$ as shown in Figure 3.16.

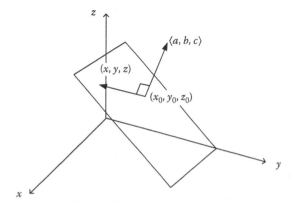

Figure 3.16 General plane in space.

So this now gives

$$\langle a, b, c \rangle \cdot \langle x - x_0, y - y_0, z - z_0 \rangle = 0$$

$$a(x - x_0) + b(y - y_0) + c(z - z_0) = 0$$

or collecting all the constants onto one side gives the *general equation of a plane* as

$$ax + by + cz = d \qquad (3.8)$$

Example 3.11

Find the equation of a plane that acts as a "mirror" for the points $(1,3,-1)$ and $(1,-1,1)$ as shown in Figure 3.17.

The plane is perpendicular to the line joining the two points.

$$\underline{n} = (1,-1,1) - (1,3,-1) = \langle 0,-4,2 \rangle$$

z

(1, –1, 1)

M

(1, 3, –1)

y

x

Figure 3.17 Plane with a mirror image of a point.

To find the midpoint M, use the midpoint formula $\left(\dfrac{x_1 + x_2}{2}, \dfrac{y_1 + y_2}{2}, \dfrac{z_1 + z_2}{2} \right)$, which gives

$$\left(\frac{1+1}{2}, \frac{3-1}{2}, \frac{-1+1}{2} \right) = (1,1,0)$$

Having a point and the normal vector, the equation of the plane is given by using Equation 3.8 as

$$0(x-1) - 4(y-1) + 2(z-0) = 0$$

$$4y - 2z = 4$$

3.3   Applications

Example 3.12: Resultant Smoke Flow

A small fire in a corridor has smoke flowing vertically at a speed of $V_s = 1.3$ ms^{-1}. An open window blows wind due east at a speed of $V_w = 2.5$ ms^{-1} as shown in Figure 3.18.

The resultant speed V_R and direction θ the smoke moves can be obtained using vectors as follows: $V_R^2 = 1.3^2 + 2.5^2 = 7.94$ and $V_R = 2.8$ ms^{-1}.

The direction of smoke flow relative to the horizontal is $\theta = \tan^{-1} \left(\dfrac{1.3}{2.5} \right) = 27.5°.$

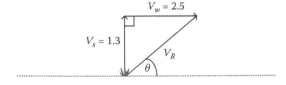

$V_w = 2.5$

$V_s = 1.3$

V_R

θ

Figure 3.18 Wind effect on fire smoke.

Example 3.13: Applicability of the Scalar Product

1. If the force $\underline{F}$ acts on a body that moves through a distance $\underline{d}$, the work done W is given by the formula

$$W = \underline{F}.\underline{d} = |\underline{F}||\underline{d}|\cos\theta \qquad (3.9)$$

Here, the $\cos\theta$ factor in the scalar product allows for the fact that it is only the component of $\underline{F}$ along the direction of motion that does the work.

2. Given two unit vectors $\hat{\underline{u}}_1$ and $\hat{\underline{u}}_2$, then the angle between these two vectors is given by

$$\cos\theta = \hat{\underline{u}}_1.\hat{\underline{u}}_2 \qquad (3.10)$$

Example 3.14: Force on a Charge in a Magnetic Field

In a region of magnetic induction given by $\underline{B}$, a particle moves with velocity $\underline{v}$ and charge q as shown in Figure 3.19. The particle experiences a force $\underline{F}$ whose size or magnitude is given by $qvB\sin\theta$ and whose direction happens to be perpendicular to the directions of both the vectors $\underline{v}$ and $\underline{B}$. So, it turns out that $\underline{F} = q\underline{v} \times \underline{B}$ gives the right magnitude and direction for the force and as a consequence the particle gets deflected.

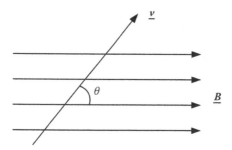

Figure 3.19 Particle moving in a magnetic field.

Problems

3.1 Find a unit vector $\hat{\underline{u}}$ in the direction of the vector $\underline{AB}$, where the points A and B are given by $A(\underline{i} + 3\underline{j} + 5\underline{k})$ and $B(3\underline{i} + 2\underline{j} + \underline{k})$.

3.2 Find the angle between the vectors $\underline{a} = \langle 1, 2, -4 \rangle$ and $\underline{b} = \langle 4, 2, 7 \rangle$.

3.3 Find the value of the constant α, which makes the vectors $\underline{a} = \langle 1, 2, \alpha \rangle$ and $\underline{b} = \langle 2, 3, 4 \rangle$ perpendicular to each other.

3.4 Figure 3.20 shows a rhombus OACB, with sides $\underline{OA}$ and $\underline{OB}$ being given by the vectors $\underline{a}$ and $\underline{b}$, respectively. Given the diagonals $\underline{OC} = \underline{a} + \underline{b}$ and $\underline{BA} = \underline{a} - \underline{b}$, show that the diagonals are perpendicular to each other.

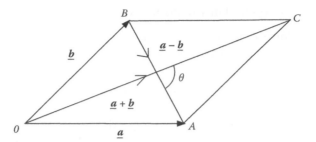

Figure 3.20 A rhombus with sides as vectors $\underline{a}$ and $\underline{b}$.

3.5 Find a vector perpendicular to both the vectors $\underline{a} = 3\underline{i} + \underline{j} - 2\underline{k}$ and $\underline{b} = \underline{i} - 2\underline{j} + \underline{k}$.

3.6 Show that the two lines given by

$$\underline{r_1} = \underline{i} + \underline{j} + 3\underline{k} + s(3\underline{i} + 8\underline{j} + 2\underline{k})$$

$$\underline{r_2} = 2\underline{i} + 5\underline{j} - \underline{k} + t(\underline{i} + 2\underline{j} + 3\underline{k})$$

intersect and find their point of intersection.

3.7 The three points $A(2\underline{i} + \underline{j} - \underline{k})$, $B(\underline{i} - 2\underline{j} + 5\underline{k})$, and $C(3\underline{i} + 2\underline{j} - 2\underline{k})$ lie in a plane. Find the equation of the plane passing through the three points.

3.8 The 2-D rotational flow of a fluid can be represented by the vector field

$$\bar{F}(x, y) = \frac{\langle -y, x \rangle}{\sqrt{x^2 + y^2}}$$

which is not defined at (0, 0).

a. Determine that the magnitude of the vector field.

b. Hence, sketch the vector field.

3.9 Smoke flows vertically upward at a speed of 0.8 ms^{-1} when it is blown by wind traveling at 0.6 ms^{-1} in a southeasterly direction. Find the resultant speed of the smoke flow and its direction relative to the vertical.

4 Determinants and Matrices

4.1 Background

Matrices with different dimensions are a standard method for solving a wide range of problems in many disciplines. In science and engineering, when considering particular problems, the derivation of the solutions to these problems can come down to solving a system of linear equations. In this chapter, different methods for solving a linear system of equations will be discussed and these methods involve the concepts of determinants and matrices.

4.2 Introduction to Determinants

Determinants arise naturally in the solution of a set of linear equations. Also they will help solve linear equations using the matrix inversion method (see Section 4.3.6). Starting with a general set of two linear simultaneous equations as follows:

$$ax + by = e \qquad\qquad (4.1)$$

$$cx + dy = f \qquad\qquad (4.2)$$

where a, b, c, d, e, and f are constants. How do you find the values of x and y? Since the values of a, b, c, and d are not known, all that can be done is try to eliminate either x or y.

If one decides to get rid of y, then to make the coefficients of y the same, first multiply Equation 4.1 by d and Equation 4.2 by b:

$$adx + bdy = de \qquad\qquad (4.3)$$

$$bcx + bdy = bf \qquad\qquad (4.4)$$

Then subtracting Equation 4.4 from Equation 4.3:

$$adx - bcx = de - bf \quad \therefore \quad x(ad - bc) = de - bf \quad \therefore \quad x = \frac{de - bf}{ad - bc}$$

A similar approach, getting rid of x, would give the result as

$$y = \frac{af - ce}{ad - bc}$$

Now provided that the denominator term $ad - bc \neq 0$, then a symbol can be used to define a 2-by-2 determinant using parallel lines (| |), as shown next:

$$\begin{vmatrix} a & b \\ c & d \end{vmatrix} \triangleq ad - bc \qquad (4.5)$$

Note: The symbol $\triangleq$ means "defined as."

This definition given by Equation 4.5 is just the difference in product of the diagonals.

Notice also how these answers for x and y contain lots of similar expressions. Using the definition of a determinant given earlier, they can be written in the following form:

$$x = \frac{\begin{vmatrix} e & b \\ f & d \end{vmatrix}}{\begin{vmatrix} a & b \\ c & d \end{vmatrix}}, \quad y = \frac{\begin{vmatrix} a & e \\ c & f \end{vmatrix}}{\begin{vmatrix} a & b \\ c & d \end{vmatrix}}$$

or alternatively as

$$\frac{x}{\begin{vmatrix} e & b \\ f & d \end{vmatrix}} = \frac{y}{\begin{vmatrix} a & e \\ c & f \end{vmatrix}} = \frac{1}{\begin{vmatrix} a & b \\ c & d \end{vmatrix}} \qquad (4.6)$$

This is called Cramer's rule. Equation 4.6 is a nice and neat way to represent the solutions to Equations 4.1 and 4.2.

4.2.1   2 × 2 Determinants

Now using the definition of a determinant given by Equation 4.5

$$\begin{vmatrix} a & b \\ c & d \end{vmatrix} \triangleq ad - bc$$

Example 4.1

Calculate the following determinant: $\begin{vmatrix} 6 & 2 \\ 3 & 4 \end{vmatrix}$

Solution:

$$\begin{vmatrix} 6 & 2 \\ 3 & 4 \end{vmatrix} = 6 \times 4 - 2 \times 3 = 24 - 6 = 18$$

Example 4.2

Calculate the following determinant: $\begin{vmatrix} -2 & -5 \\ 3 & 4 \end{vmatrix}$

Solution:

$$\begin{vmatrix} -2 & -5 \\ 3 & 4 \end{vmatrix} = -2 \times 4 - (-5) \times 3 = -8 + 15 = 7$$

4.2.2   Properties of Determinants

Looking at these answers in the above examples, one can notice some facts about determinants.

- If rows are swapped for columns, the value of the determinant is unchanged.

$$\begin{vmatrix} a & b \\ c & d \end{vmatrix} = \begin{vmatrix} a & c \\ b & d \end{vmatrix}$$

- If one row (or column) is equal to the other, the value of the determinant is zero.

$$\begin{vmatrix} a & b \\ a & b \end{vmatrix} = 0 = \begin{vmatrix} a & a \\ c & c \end{vmatrix}$$

- If one row (or column) is a multiple of the other, the value of the determinant is zero.

$$\begin{vmatrix} a & b \\ ka & kb \end{vmatrix} = 0 = \begin{vmatrix} a & ka \\ c & kc \end{vmatrix}$$

This last property shows that, for example,

$$\begin{vmatrix} 100 & 500 \\ 1 & 5 \end{vmatrix} = 0$$

4.2.2.1 Multiplying a Determinant by a Number

Look what happens if every element of a 2×2 determinant is multiplied by 3:

$$\begin{vmatrix} 3p & 3r \\ 3q & 3s \end{vmatrix} = 9ps - 9qr = 9(ps - pr) = 9 \begin{vmatrix} p & r \\ q & s \end{vmatrix}$$

In other words, the value of the determinant is multiplied by 9.

Multiplying just one row (or column) of a determinant by a number has the effect of multiplying the value of the determinant by that number. For example,

$$5 \begin{vmatrix} m & n \\ o & p \end{vmatrix} = \begin{vmatrix} 5m & n \\ 5o & p \end{vmatrix} = \begin{vmatrix} m & n \\ 5o & 5p \end{vmatrix} = 5(mp - on)$$

It can be seen why the last property mentioned earlier comes about:

$$\begin{vmatrix} 100 & 500 \\ 1 & 5 \end{vmatrix} = 100 \begin{vmatrix} 1 & 5 \\ 1 & 5 \end{vmatrix} = 100(5 - 5) = 100(0) = 0$$

In fact, there is a much "stronger" property that can be easily proved. First, a numerical example to show this:

$$\begin{vmatrix} 102 & 504 \\ 1 & 5 \end{vmatrix} = \begin{vmatrix} 2 & 4 \\ 1 & 5 \end{vmatrix} = 10 - 4 = 6$$

Here the determinant is made much easier to work out by subtracting 100 and 500 from the top row.

This can be made a general rule as follows:

$$\begin{vmatrix} a+kc & b+kd \\ c & d \end{vmatrix} = (a+kc)d - (b+kd)c = ad + kcd - bc - kcd$$

$$= ad - bc = \begin{vmatrix} a & b \\ c & d \end{vmatrix}$$

Or the value of a determinant does not change if one adds to one row a multiple of another row.

This might not seem important for 2×2 determinants, but all the preceding rules apply to determinants of any size, and it is with larger ones that they come in useful.

4.2.3 3×3 Determinants

The basic building block of a determinant is the 2×2 one already discussed. All larger ones are broken down into combinations of 2×2 determinants, according to the rule of signs given as

$$\begin{vmatrix} + & - & + & \dots \\ - & + & - & \\ + & - & + & \\ \vdots & & & \end{vmatrix}$$

Then going along any row or column, multiplying each element by

1. The sign in that place

2. The 2×2 determinant revealed when covering up the row and column with that element in

This following example shows how this is carried in practice.

Example 4.3

Take the determinant

$$\begin{vmatrix} 3 & 4 & 2 \\ 0 & 1 & 5 \\ 1 & 6 & 8 \end{vmatrix}$$

Suppose this is expanded along the top row. The element 3 is multiplied by "+," and by the 2×2 determinant $\begin{vmatrix} 1 & 5 \\ 6 & 8 \end{vmatrix}$, since that is what is revealed if one covers up the row and column containing the element 3. Next move along to the element 4. This is multiplied by "−" and by $\begin{vmatrix} 0 & 5 \\ 1 & 8 \end{vmatrix}$. Finally, take

the element 2. This is multiplied by "+" and by $\begin{vmatrix} 0 & 1 \\ 1 & 6 \end{vmatrix}$. Now adding up the

three parts gives the value of the determinant as

$$\begin{vmatrix} 3 & 4 & 2 \\ 0 & 1 & 5 \\ 1 & 6 & 8 \end{vmatrix} = +3 \begin{vmatrix} 1 & 5 \\ 6 & 8 \end{vmatrix} - 4 \begin{vmatrix} 0 & 5 \\ 1 & 8 \end{vmatrix} + 2 \begin{vmatrix} 0 & 1 \\ 1 & 6 \end{vmatrix}$$

$$= 3(-22) - 4(-5) + 2(-1) = -48$$

Suppose trying to expand this using a different row; let's say row 2. Then the working is similar, but remember that now the expansion starts with a minus sign this time giving

$$\begin{vmatrix} 3 & 4 & 2 \\ 0 & 1 & 5 \\ 1 & 6 & 8 \end{vmatrix} = -0 \begin{vmatrix} 4 & 2 \\ 6 & 8 \end{vmatrix} + 1 \begin{vmatrix} 3 & 2 \\ 1 & 8 \end{vmatrix} - 5 \begin{vmatrix} 3 & 4 \\ 1 & 6 \end{vmatrix}$$

$$= 0(\text{not necessary}) + 1(22) - 5(14) = -48$$

The same result is obtained expanding along any row or column.

Now what about using a column instead of a row? From the first property, the value of the determinant is unchanged if the rows and columns are swapped. Therefore, a column can be regarded as equivalent to a row, and the calculation can be done in a similar way, for example, using the last column:

$$\begin{vmatrix} 3 & 4 & 2 \\ 0 & 1 & 5 \\ 1 & 6 & 8 \end{vmatrix} = +2 \begin{vmatrix} 0 & 1 \\ 1 & 6 \end{vmatrix} - 5 \begin{vmatrix} 3 & 4 \\ 1 & 6 \end{vmatrix} + 8 \begin{vmatrix} 3 & 4 \\ 0 & 1 \end{vmatrix}$$

$$= 2(-1) - 5(14) + 8(3) = -48$$

Why all the fuss about being able to use any row or column to evaluate a determinant? Because, as seen, it is very handy if some of the elements are zero. Thus, using the first column or the second row would be quicker ways of evaluating the above determinant.

Example 4.4

Evaluate the following determinants:

1. $\begin{vmatrix} 3 & 6 & 2 \\ 5 & 2 & 1 \\ 0 & 4 & 0 \end{vmatrix}$ Expanding along row 3 would speed the working.

2.
$$\begin{vmatrix} -4 & 2 & 7 \\ 3 & 5 & 0 \\ 1 & 8 & 0 \end{vmatrix}$$
Expanding along column 3 would help.

Example 4.5

Find the values of k for which

$$\begin{vmatrix} k & 3 & -5 \\ 0 & k-1 & 4 \\ 0 & 0 & k+5 \end{vmatrix} = 0$$

Solution: Expanding the determinant along first column gives

$$\begin{vmatrix} k & 3 & -5 \\ 0 & k-1 & 4 \\ 0 & 0 & k+5 \end{vmatrix} = k \begin{vmatrix} k-1 & 4 \\ 0 & k+5 \end{vmatrix} = k(k-1)(k+5) = 0$$

This implies $k = 0$, $k = 1$, or $k = -5$ as the solutions.

4.3   Introduction to Matrices

In English, one thinks of the word *matrix* as meaning a "grid" or an array." In mathematics, it has a special meaning: An "object" made up of a rectangular arrangement of "things," or *elements*, which behaves according to certain rules. These elements might be numbers, letters, complex numbers, or other things, but the rules that govern matrix arithmetic are always the same.

When a matrix is written, it is always enclosed in brackets (either curly or square). This is very important, because there are other rectangular arrays, as seen in the previous section, called determinants that are written differently, and the different types must not be confused!

Example 4.6

All the following are matrices

$$\begin{pmatrix} 1 & 0 \\ 3 & 2 \end{pmatrix}, \qquad \begin{pmatrix} a \\ b \\ c \\ d \end{pmatrix}, \qquad \begin{pmatrix} 2+j & -5-3j \\ j & 1-j \end{pmatrix}$$

Note: Matrices that have a single column (as the second one above) are called column vectors and those that have a single row are called row vectors.

4.3.1   Order of a Matrix

Clearly, the preceding matrices have different sizes or "order." The order of a matrix is the number of rows by the number of columns it has and is written generally as m × n, where m is the number of rows and n is the number of columns.

Example 4.7

In Example 4.6, the order of the matrices are as follows: 2 × 2, 4 × 1, and 2 × 2.

If two matrices are equal, clearly they must be the same size and shape; also corresponding elements must be identical. So if, for example, the two matrices $\begin{pmatrix} a & b \\ c & d \end{pmatrix}$ and $\begin{pmatrix} 3 & -5 \\ 2 & 1 \end{pmatrix}$ are equal, then there are four equations that must be true: $a = 3$, $b = -5$, $c = 2$, and $d = 1$.

4.3.2   Addition and Subtraction

If it is required to add or subtract two matrices, to be able to do this, they first have to have the same size and shape, that is, have the same order. Then it is a matter of simply adding or subtracting the corresponding elements as shown in the next example.

Example 4.8

$$\begin{pmatrix} 2 & 1 & 8 \\ 1 & 5 & 3 \\ 0 & 6 & 4 \end{pmatrix} + \begin{pmatrix} 6 & 1 & 8 \\ 0 & 2 & 1 \\ 5 & 3 & 0 \end{pmatrix} = \begin{pmatrix} 8 & 2 & 16 \\ 1 & 7 & 4 \\ 5 & 9 & 4 \end{pmatrix}$$

or

$$\begin{pmatrix} 2 & 1 & 8 \\ 1 & 5 & 3 \\ 0 & 6 & 4 \end{pmatrix} - \begin{pmatrix} 6 & 1 & 8 \\ 0 & 2 & 1 \\ 5 & 3 & 0 \end{pmatrix} = \begin{pmatrix} -4 & 0 & 0 \\ 1 & 3 & 2 \\ -5 & 3 & 4 \end{pmatrix}$$

4.3.3   Matrix Multiplication

4.3.3.1   Multiplying a Matrix by a Scalar

Adding a matrix A to itself, it makes sense to call the result 2A. The effect will be to multiply each element by 2 as follows:

$$\begin{pmatrix} 3 & 6 \\ 4 & 5 \end{pmatrix} + \begin{pmatrix} 3 & 6 \\ 4 & 5 \end{pmatrix} = \begin{pmatrix} 6 & 12 \\ 8 & 10 \end{pmatrix} = 2\begin{pmatrix} 3 & 6 \\ 4 & 5 \end{pmatrix}$$

This idea can be extended to apply to multiplying a matrix by any number (scalar): Simply multiply each element by that number as shown next:

$$k \begin{pmatrix} a & b \\ c & d \end{pmatrix} = \begin{pmatrix} ka & kb \\ kc & kd \end{pmatrix}$$

Note: This is different than multiplying a determinant by a constant.

4.3.3.2 Multiplying Two Matrices

In order to discuss this, it is better to define the way to describe the size and shape (or order) of a matrix. It takes too long to say: "A matrix with three rows and four columns." Instead this can be referred to as a "3 × 4 matrix." When looking at a matrix, first go

down the left-hand side:
$$3 \begin{pmatrix} 1 & 2 & -3 & 7 \\ 3 & 5 & 1 & -2 \\ 0 & 2 & 6 & 8 \end{pmatrix} \quad \text{This is a } \mathbf{3 \times 4 \text{ matrix}}$$

4

then along the bottom

4.3.3.3 How to Multiply Two Matrices

Consider a matrix A order m × n and a second matrix B of order p × q, then the multiplication of the two matrices, AB, is only defined if the number of columns of A, given by (n), is equal to the number of rows of B, given by (p), that is (n = p the inner numbers). Then the product exists and is of order given by m × q (the outer numbers). This can be represented as follows:

$$\text{If } A = \begin{pmatrix} 1 & -2 & 7 \\ 2 & 5 & 4 \end{pmatrix} \quad \text{and} \quad B = \begin{pmatrix} 2 & 6 \\ 1 & 8 \\ 3 & 5 \end{pmatrix}, \text{ then how to determine the product } AB,$$

$$2 \times 3 \qquad\qquad 3 \times 2$$

Now since inner numbers are the same 3 = 3, that is, n = p, for these two matrices, then this product exists. The answer will be a matrix of order given by the outer numbers, which is 2 × 2. This can be represented at the moment generally as

$$\begin{pmatrix} 1 & -2 & 7 \\ 2 & 5 & 4 \end{pmatrix} \begin{pmatrix} 2 & 6 \\ 1 & 8 \\ 3 & 5 \end{pmatrix} = \begin{pmatrix} R_1C_1 & R_1C_2 \\ R_2C_1 & R_2C_2 \end{pmatrix}$$

$$2 \times 3 \qquad 3 \times 2 \qquad\qquad 2 \times 2$$

This product matrix has four elements in it. These elements are calculated by seeing the location of each element in the product matrix and carrying out the corresponding multiplication of the row and column from the original two matrices and adding all the multiplications together.

In the above example, the first element in the product matrix, R_1C_1 is given by row 1 of the first matrix multiplying column 1 of the second matrix as follows:

$$R_1C_1 = (1 - 2 \ 7) \begin{pmatrix} 2 \\ 1 \\ 3 \end{pmatrix} = (1) \times (2) + (-2) \times (1) + (7) \times (3) = 2 - 2 + 21 = 21$$

Note: Here the multiplication is carried out with corresponding elements, that is, the first element of the row with the first element of the column, and so on.

Similarly, for the other three elements of the product matrix:

$$R_1C_2 = (1 \ -2 \ 7) \begin{pmatrix} 6 \\ 8 \\ 5 \end{pmatrix} = (1) \times (6) + (-2) \times (8) + (7) \times (5) = 6 - 16 + 35 = 25$$

$$R_2C_1 = (2 \ 5 \ 4) \begin{pmatrix} 2 \\ 1 \\ 3 \end{pmatrix} = (2) \times (2) + (5) \times (1) + (4) \times (3) = 4 + 5 + 12 = 21$$

$$R_2C_2 = (2 \ 5 \ 4) \begin{pmatrix} 6 \\ 8 \\ 5 \end{pmatrix} = (2) \times (6) + (5) \times (8) + (4) \times (5) = 12 + 40 + 20 = 72$$

Now these four elements can be put into the result of the product as

$$\begin{pmatrix} 1 & -2 & 7 \\ 2 & 5 & 4 \end{pmatrix} \begin{pmatrix} 2 & 6 \\ 1 & 8 \\ 3 & 5 \end{pmatrix} = \begin{pmatrix} 21 & 25 \\ 21 & 72 \end{pmatrix}$$

Finally,

$$AB = \begin{pmatrix} 21 & 25 \\ 21 & 72 \end{pmatrix}$$

Example 4.9

Given the following matrices,

$$A = \begin{pmatrix} 50 & 100 & 150 & 50 \\ 60 & 80 & 25 & 50 \end{pmatrix} \text{ and } B = \begin{pmatrix} 1.5 \\ 2.0 \\ 3.5 \\ 4.0 \end{pmatrix}, \text{ calculate the product } AB.$$

Solution: Matrix A has order 2×4 and matrix B has order 4×1. Since the inner numbers match, then the product exists and the result will be a matrix of order 2×1, that is, the outer numbers as follows:

$$\begin{pmatrix} 50 & 100 & 150 & 50 \\ 60 & 80 & 25 & 50 \end{pmatrix} \begin{pmatrix} 1.5 \\ 2.0 \\ 3.5 \\ 4.0 \end{pmatrix} = \begin{pmatrix} R_1C_1 \\ R_2C_1 \end{pmatrix} = \begin{pmatrix} 1000 \\ 537.5 \end{pmatrix}$$

This is the essence of matrix multiplication. Note that in this last example the orders of the matrices were

$$(2 \times 4)\,(4 \times 1) \text{ giving a } (2 \times 1) \text{ matrix}$$

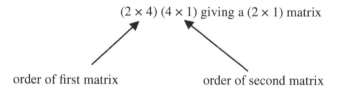

order of first matrix   order of second matrix

and that the multiplication could only be done because the "4's matched": the number of columns in the first one matched the number of rows in the second one.

Could these particular matrices have been multiplied the other way around, that is, does the product BA exist? By considering the structure this gives

$$\begin{pmatrix} 1.5 \\ 2.0 \\ 3.5 \\ 4.0 \end{pmatrix} \begin{pmatrix} 50 & 100 & 150 & 50 \\ 60 & 80 & 25 & 50 \end{pmatrix} = ?$$

Because these numbers don't match up, implies can't do that matrix multiplication this way round: *the product doesn't exist.*

(4×1) $(2) \times 4$

Example 4.10

$$\begin{pmatrix} 3 & 5 \\ 2 & 1 \end{pmatrix} \begin{pmatrix} 1 & 0 \\ 4 & 2 \end{pmatrix} = \begin{pmatrix} 23 & 10 \\ 6 & 2 \end{pmatrix}$$

Example 4.11

$$\begin{pmatrix} 2 & 3 \\ 1 & 4 \end{pmatrix} \begin{pmatrix} 0 & 4 & 2 \\ 2 & 5 & 1 \end{pmatrix} = \begin{pmatrix} 6 & 23 & 7 \\ 8 & 24 & 6 \end{pmatrix}$$

Example 4.12

$$\begin{pmatrix} 3 \\ 6 \\ 1 \end{pmatrix} (6 \quad 2 \quad 8) = \begin{pmatrix} 18 & 6 & 24 \\ 36 & 12 & 48 \\ 6 & 2 & 8 \end{pmatrix}$$

What is observed is *even if both matrix products AB and BA exist, they may not be equal.* In general, for matrices, $AB \neq BA$.

Note: That this is quite different behaviour from ordinary numbers!

4.3.4 Special Matrices

As with normal algebra, the number 1 has the property that $1 \times a = a \times 1 = a$. With matrices, there is a similar situation using the identity matrix I as follows:

$$I_2 = \begin{pmatrix} 1 & 0 \\ 0 & 1 \end{pmatrix} \text{ and } I_3 = \begin{pmatrix} 1 & 0 & 0 \\ 0 & 1 & 0 \\ 0 & 0 & 1 \end{pmatrix} \text{ etc.}$$

These are called the *identity matrix*, I, which have the special property that for any other matrix A gives $A \, I = I \, A = A$. The identity matrix is always square, but can be any size, such as 2×2, 3×3, 4×4, and so forth.

Finally, this section will conclude with other special types of matrices:

- A *square matrix* could have all its elements equal to zero: this is called the *null matrix*.

- If all the elements except those on the main diagonal are zero, then it is called a *diagonal matrix*.

- As noted above, the *identity matrix*, I, is a diagonal matrix with all its nonzero elements equal to 1.

- The *transpose* of a matrix is a new matrix obtained by swopping rows into columns and vice versa. To transpose a matrix, row 1 will become column 1, and row 2 will become column 2, and so forth. The transpose of matrix A is written as A^T. For example, if

$$A = \begin{pmatrix} 1 & 0 & 5 \\ 4 & 2 & 1 \end{pmatrix}$$

then

$$A^T = \begin{pmatrix} 1 & 4 \\ 0 & 2 \\ 5 & 1 \end{pmatrix}$$

and it can be seen that the transpose of a 2×3 matrix must be a 3×2 matrix. The transpose of a square matrix will also be a square matrix, of the same order. What do you notice about these two matrices and their transposes?

$$B = \begin{pmatrix} 3 & 1 & 8 \\ 1 & 2 & 4 \\ 8 & 4 & 9 \end{pmatrix} \quad B^T = \begin{pmatrix} 3 & 1 & 8 \\ 1 & 2 & 4 \\ 8 & 4 & 9 \end{pmatrix}$$

$$C = \begin{pmatrix} 0 & -1 & -2 \\ 1 & 0 & 5 \\ 2 & -5 & 0 \end{pmatrix} \quad C^T = \begin{pmatrix} 0 & 1 & 2 \\ -1 & 0 & -5 \\ -2 & 5 & 0 \end{pmatrix}$$

B is an example of a *symmetric matrix*: $B = B^T$.
C is an example of a *skew-symmetric matrix*: $C = -C^T$.

4.3.5   Powers of Matrices

As with normal algebraic expressions, powers of matrices are calculated as follows:

$$A^2 = A \times A$$
$$A^3 = A \times A \times A$$
$$A^4 = A \times A \times A \times A$$
$$\vdots$$

Note: $A^3 = A^2 \times A$, $A^4 = A^2 \times A^2$, *and so forth.*

Example 4.13

Given a square matrix $A = \begin{pmatrix} 3 & 5 \\ 2 & 1 \end{pmatrix}$, calculate A^2.

Solution:

$$A^2 = A \times A = \begin{pmatrix} 3 & 5 \\ 2 & 1 \end{pmatrix} \times \begin{pmatrix} 3 & 5 \\ 2 & 1 \end{pmatrix} = \begin{pmatrix} 19 & 20 \\ 8 & 11 \end{pmatrix}$$

4.3.6   Inverse of a Square Matrix

The inverse of a matrix A is denoted by A^{-1} such that $AA^{-1} = I$, where I is the identity matrix.

To calculate the inverse of a square matrix A use the following formula:

$$A^{-1} = \frac{1}{|A|} A_c^T \tag{4.7}$$

where $|A|$ is the determinant of the matrix A, and A_c^T is the matrix of cofactors of A transposed. The cofactors of the matrix A are given by first taking into account an appropriate sign for each element in the matrix using the following structure:

$$\begin{pmatrix} + & - & + & \dots \\ - & + & - \\ + & - & + \\ \vdots \end{pmatrix}$$

Then each element is made up of the sign with the determinant of elements that are left after removing the row and column containing that element.

Generally, if A is a square matrix, then the "minor" of entry a_{ij} is denoted by M_{ij} and is defined to be the determinant of the submatrix that remains after the ith row and jth column are deleted from A.

The number $(-1)^{i+j} M_{ij}$ is denoted by C_{ij} and is called the cofactor of entry a_{ij}. The next examples show how this works for a 2×2 and 3×3 matrix.

Example 4.14

The determinant of a general 2×2 matrix

$$A = \begin{pmatrix} a & b \\ c & d \end{pmatrix}$$

is given as $|A| = ad - bc$ and the matrix of cofactors of

$$A_c = \begin{pmatrix} d & -c \\ -b & a \end{pmatrix} \text{ and } A_c^T = \begin{pmatrix} d & -b \\ -c & a \end{pmatrix}.$$

This gives the inverse of A as

$$A^{-1} = \frac{1}{ad - bc} \begin{pmatrix} d & -b \\ -c & a \end{pmatrix} \tag{4.8}$$

Example 4.15

Calculate the inverse of the matrix $A = \begin{pmatrix} 2 & 3 \\ 1 & 5 \end{pmatrix}$.

Solution:

$$|A| = 10 - 3 = 7$$

$$A^{-1} = \frac{1}{7} \begin{pmatrix} 5 & -3 \\ -1 & 2 \end{pmatrix}.$$

Check

$$AA^{-1} = \frac{1}{7} \begin{pmatrix} 7 & 0 \\ 0 & 7 \end{pmatrix} = \begin{pmatrix} 1 & 0 \\ 0 & 1 \end{pmatrix} = I$$

Example 4.16

Following is an example of the inverse of a 3×3 matrix using the formula $A^{-1} = \frac{1}{|A|} A_c^T$.

Find the inverse of the matrix

$$A = \begin{pmatrix} 0 & 0 & 1 \\ 2 & -1 & 3 \\ 1 & 1 & 4 \end{pmatrix}$$

Solution: Expanding along row 1 gives the determinant as

$$|A| = 1 \begin{vmatrix} 2 & -1 \\ 1 & 1 \end{vmatrix} = 3.$$

Matrix of cofactors is

$$A_c = \begin{pmatrix} +\begin{vmatrix} -1 & 3 \\ 1 & 4 \end{vmatrix} & -\begin{vmatrix} 2 & 3 \\ 1 & 4 \end{vmatrix} & +\begin{vmatrix} 2 & -1 \\ 1 & 1 \end{vmatrix} \\ -\begin{vmatrix} 0 & 1 \\ 1 & 4 \end{vmatrix} & +\begin{vmatrix} 0 & 1 \\ 1 & 4 \end{vmatrix} & -\begin{vmatrix} 0 & 0 \\ 1 & 1 \end{vmatrix} \\ +\begin{vmatrix} 0 & 1 \\ -1 & 3 \end{vmatrix} & -\begin{vmatrix} 0 & 1 \\ 2 & 3 \end{vmatrix} & +\begin{vmatrix} 0 & 0 \\ 2 & -1 \end{vmatrix} \end{pmatrix}$$

$$A_c = \begin{pmatrix} -7 & -5 & 3 \\ 1 & -1 & 0 \\ 1 & 2 & 0 \end{pmatrix}$$

$$A_c^T = \begin{pmatrix} -7 & 1 & 1 \\ -5 & -1 & 2 \\ 3 & 0 & 0 \end{pmatrix}$$

So therefore, $A^{-1} = \dfrac{1}{3} \begin{pmatrix} -7 & 1 & 1 \\ -5 & -1 & 2 \\ 3 & 0 & 0 \end{pmatrix}$.

Check to see that $AA^{-1} = I$.

Note: In general, for higher-order matrices, for example, 3 × 3 and 4 × 4, finding inverses can be difficult and as such it is much easier to use a matrix calculator to work them out.

4.3.7   Eigenvalues and Eigenvectors

If a column matrix is regarded as representing a vector, then premultiplying this column matrix by a square matrix will result in another column matrix or another vector, for example,

$$\begin{pmatrix} 2 & 1 \\ -2 & 5 \end{pmatrix}\begin{pmatrix} 6 \\ 3 \end{pmatrix} = \begin{pmatrix} 15 \\ 3 \end{pmatrix}$$

One could say the vector $\begin{pmatrix} 6 \\ 3 \end{pmatrix}$ has been transformed into the vector $\begin{pmatrix} 15 \\ 3 \end{pmatrix}$

by the square matrix $\begin{pmatrix} 2 & 1 \\ -2 & 5 \end{pmatrix}$.

Both the direction and the length, or magnitude (see Chapter 3, Section 3.1.1), of the vector have changed. But will this be true for every vector that gets transformed with this matrix?

Clearly, all the matrix multiplications cannot be done by hand. An Excel workbook could easily be created that shows both vectors and changes the direction of the "input" vector and the modulus. It is noticed that there are two directions for which the vectors seem to be "in line with one another," in other words the modulus may have changed as a result of the transformation but not the direction.

If the input vector is any vector along the direction $\begin{pmatrix} 1 \\ 1 \end{pmatrix}$, let's call it

$\begin{pmatrix} k \\ k \end{pmatrix}$, where $0 \neq k \in \mathbb{R}$, then the "output" vector is also along the same direction.

Moreover, for any such vector, the output vector is always 3 times as long as the input, as follows:

$$\begin{pmatrix} 2 & 1 \\ -2 & 5 \end{pmatrix}\begin{pmatrix} k \\ k \end{pmatrix} = \begin{pmatrix} 3k \\ 3k \end{pmatrix} = 3\begin{pmatrix} k \\ k \end{pmatrix}$$

Also, if the input vector is any vector along the direction $\begin{pmatrix} 1 \\ 2 \end{pmatrix}$, let's call it

$\begin{pmatrix} k \\ 2k \end{pmatrix}$, where $0 \neq k \in \mathbb{R}$, then the output vector is also along the same direction.

Moreover, for such vectors, the output vector is always 4 times as long as the input, for

$$\begin{pmatrix} 2 & 1 \\ -2 & 5 \end{pmatrix}\begin{pmatrix} k \\ 2k \end{pmatrix} = \begin{pmatrix} 4k \\ 8k \end{pmatrix} = 4\begin{pmatrix} k \\ 2k \end{pmatrix}$$

These two "multiplying factors," 3 and 4, are called the *eigenvalues* of the matrix $\begin{pmatrix} 2 & 1 \\ -2 & 5 \end{pmatrix}$.

Each eigenvalue is associated with a direction; this direction is called the *eigenvector* associated with that eigenvalue. (*Eigen* is German for "characteristic.")

Hence the matrix $\begin{pmatrix} 2 & 1 \\ -2 & 5 \end{pmatrix}$ has an eigenvalue 3, with associated eigenvec-

tor $\begin{pmatrix} k \\ k \end{pmatrix}$, $0 \neq k \in \mathbb{R}$; and an eigenvalue 4 with associated eigenvector $\begin{pmatrix} k \\ 2k \end{pmatrix}$,

$0 \neq k \in \mathbb{R}$.

Eigenvalues and eigenvectors are important properties of matrices, with many applications in engineering as well as in pure and applied mathematics.

One can determine whether a square matrix has eigenvalues and eigenvectors, and if so, what they are. The deciding property seems to be that the output vector is a multiple of the input vector. Taking the matrix from earlier, this can be written as

$$\begin{pmatrix} 2 & 1 \\ -2 & 5 \end{pmatrix} \begin{pmatrix} x \\ y \end{pmatrix} = \lambda \begin{pmatrix} x \\ y \end{pmatrix}$$

where $\lambda \in \mathbb{R}$ is the multiple or

$$AX = \lambda X$$
$$\therefore \lambda X - AX = 0$$
$$\therefore (\lambda I - A)X = 0$$

Note: This could also have been written as $(A - \lambda I)X = 0.$

Here, in matrix form, is a set of two homogeneous equations in two variables. There will always be the trivial solution, $x = y = 0$, but for a nontrivial solution to exist,

$$\det(\lambda I - A) = 0,$$

In the preceding example,

$$\det\left[\begin{pmatrix} \lambda & 0 \\ 0 & \lambda \end{pmatrix} - \begin{pmatrix} 2 & 1 \\ -2 & 5 \end{pmatrix}\right] = \begin{vmatrix} \lambda - 2 & 1 \\ 2 & \lambda - 5 \end{vmatrix} = (\lambda^2 - 7\lambda + 12) = 0$$

$$\therefore (\lambda - 3)(\lambda - 4) = 0$$

so $\lambda = 3$ or 4.

For any square matrix A, the expression det $(\lambda I - A)$ is called the *characteristic polynomial* of A. The characteristic polynomial is sometimes written simply as $h(\lambda)$. The equation det $(\lambda I - A) = 0$, is called the *characteristic equation* of A. Then any λ with the property, A X = λ X is called an eigenvalue of A and then X is called the eigenvector of A corresponding to λ.

Example 4.17

Find the eigenvalues and associated eigenvectors of the matrix

$$A = \begin{pmatrix} 1 & 0 & -1 \\ 1 & 2 & 1 \\ 2 & 2 & 3 \end{pmatrix}$$

Solution: The characteristic equation is given by det $(\lambda I - A) = 0$. In this case

$$\begin{vmatrix} \lambda-1 & 0 & 1 \\ -1 & \lambda-2 & -1 \\ -2 & -2 & \lambda-3 \end{vmatrix} = 0$$

Evaluating along the top row

$$(\lambda-1)\{(\lambda-2)(\lambda-3)-2\}+\{2+2(\lambda-2)\} = 0$$
$$\therefore \quad (\lambda-1)\{\lambda^2 - 5\lambda + 4\} + 2(\lambda-1) = 0$$
$$\therefore \quad (\lambda-1)\{\lambda^2 - 5\lambda + 6\} = 0$$
$$\therefore \quad (\lambda-1)(\lambda-2)(\lambda-3) = 0$$
$$\therefore \quad \lambda = 1, 2, \text{ or } 3$$

There are three eigenvalues, namely 1, 2, and 3.
Take each in turn and find an associated eigenvector.

$$\text{using } \lambda = 1 \text{ in } AX = \lambda X: \quad \begin{pmatrix} 1 & 0 & -1 \\ 1 & 2 & 1 \\ 2 & 2 & 3 \end{pmatrix} \begin{pmatrix} x \\ y \\ z \end{pmatrix} = 1 \begin{pmatrix} x \\ y \\ z \end{pmatrix}$$

$$\therefore \qquad x - z = x, \qquad \qquad \text{so} \qquad z = 0$$
$$\text{Also,} \quad x + 2y + z = y \qquad \qquad \text{so} \qquad x = -y$$

So, an eigenvector could be, $\begin{pmatrix} k \\ -k \\ 0 \end{pmatrix}$, for any real (non-zero) number k.

Sometimes the "scaling factor:" k is omitted, and the eigenvector is given as $\begin{pmatrix} 1 \\ -1 \\ 0 \end{pmatrix}$, where it is understood that any multiple of this vector will also work.

Similarly,

using $\lambda = 2$ in $AX = \lambda X$: $\begin{pmatrix} 1 & 0 & -1 \\ 1 & 2 & 1 \\ 2 & 2 & 3 \end{pmatrix} \begin{pmatrix} x \\ y \\ z \end{pmatrix} = 2 \begin{pmatrix} x \\ y \\ z \end{pmatrix}$

$\therefore$ $x - z = 2x,$ so $x = -2$

also, $x + 2y + z = 2y,$ so $2y = 2y$

and $2x + 2y + 3z = 2z,$ so $2y = -z - 2x = -z + 2z = z$

So an eigenvector could be, $\begin{pmatrix} -2k \\ k \\ 2k \end{pmatrix}$, for any real (nonzero) number k.

And finally,

using $\lambda = 3$ in $AX = \lambda X$: $\begin{pmatrix} 1 & 0 & -1 \\ 1 & 2 & 1 \\ 2 & 2 & 3 \end{pmatrix} \begin{pmatrix} x \\ y \\ z \end{pmatrix} = 3 \begin{pmatrix} x \\ y \\ z \end{pmatrix}$

So an eigenvector could be, $\begin{pmatrix} k \\ -k \\ -2k \end{pmatrix}$, for any real (nonzero) number k.

Therefore, the matrix

$$A = \begin{pmatrix} 1 & 0 & -1 \\ 1 & 2 & 1 \\ 2 & 2 & 3 \end{pmatrix}$$

has eigenvalues 1, 2, and 3 with corresponding eigenvectors $\begin{pmatrix} 1 \\ -1 \\ 0 \end{pmatrix}$, $\begin{pmatrix} -2 \\ 1 \\ 2 \end{pmatrix}$,

and $\begin{pmatrix} 1 \\ -1 \\ -2 \end{pmatrix}$.

4.3.8   Diagonal Factorization of Matrices

For an eigenvalue equation $AX = \lambda X$, suppose there are n independent eigenvectors of the matrix A. If these eigenvectors are put in a column P, that is,

$$P = \begin{bmatrix} \vdots & \vdots & & & \vdots \\ X_1 & X_2 & \cdots & \cdots & X_n \\ \vdots & \vdots & & & \vdots \end{bmatrix}$$

then

$$AP = A \begin{bmatrix} \vdots & \vdots & & \vdots \\ X_1 & X_2 & \cdots \cdots & X_n \\ \vdots & \vdots & & \vdots \end{bmatrix} = \begin{bmatrix} \vdots & \vdots & & \vdots \\ \lambda_1 X_1 & \lambda_2 X_2 & \cdots \cdots & \lambda_n X_n \\ \vdots & \vdots & & \vdots \end{bmatrix}$$

$$AP = \begin{bmatrix} \vdots & \vdots & & \vdots \\ X_1 & X_2 & \cdots \cdots & X_n \\ \vdots & \vdots & & \vdots \end{bmatrix} \begin{bmatrix} \lambda_1 & \cdots & 0 \\ \vdots & \ddots & \vdots \\ 0 & \cdots & \lambda_n \end{bmatrix}$$

Diagonal eigenvalue
matrix D

Therefore, $AP = PD$ and $P^{-1}AP = D$ provided P is invertible, requiring n independent eigenvectors.

Therefore, giving the following results,

$$D = P^{-1}AP \tag{4.9}$$

$$A = PDP^{-1} \tag{4.10}$$

How do you make use of this information? Since

$$A = PDP^{-1}$$
$$A^2 = PDP^{-1}PDP^{-1} = PD^2P^{-1}$$
$$A^3 = PDP^{-1}PDP^{-1}PDP^{-1} = PD^3P^{-1}$$

and so on, this is generalized to power k:

$$\vdots$$
$$A^k = PDP^{-1}PDP^{-1}PDP^{-1}...PDP^{-1} = PD^kP^{-1}$$

This gives an important result:

$$A^k = PD^kP^{-1} \tag{4.11}$$

So, eigenvalues and eigenvectors give an effective way of finding *powers of matrices.*

The diagonal matrix D and its powers are just the powers of the diagonal elements and so any power of a matrix can be raised by just three matrices multiplied together.

The matrix A is sure to have n independent eigenvectors and therefore be diagonalizable if

- All the λ's are different or distinct, that is, no repeated eigenvalues.

- Having repeated eigenvalues means there may or may not have been independent eigenvectors.

Example 4.18

Determine the diagonal factorization of a 2×2 matrix given by

$$A = \begin{bmatrix} 4 & 1 \\ -8 & -5 \end{bmatrix}$$

Solution: A can be diagonalized if A has two independent eigenvectors. First, finding the eigenvalues and eigenvectors of A:

$$AX = \lambda X$$

$$(A - \lambda I)X = 0$$

$$\det(A - \lambda I) = 0$$

$$\det \begin{bmatrix} 4 - \lambda & 1 \\ -8 & -5 - \lambda \end{bmatrix} = 0$$

$$\lambda^2 + \lambda - 12 = 0$$

$$(\lambda + 4)(\lambda - 3) = 0$$

gives $\lambda_1 = -4$ and $\lambda_2 = 3$.
The corresponding eigenvectors are for

$$\lambda_1 = 3, \text{ the eigenvector is } X_1 = \begin{bmatrix} 1 \\ -1 \end{bmatrix}.$$

$$\lambda_2 = -4, \text{ the eigenvector is } X_2 = \begin{bmatrix} 1 \\ -8 \end{bmatrix}.$$

So,

$$P = \begin{bmatrix} 1 & 1 \\ -1 & -8 \end{bmatrix}$$

which gives

$$P^{-1} = -\frac{1}{7}\begin{bmatrix} -8 & -1 \\ 1 & 1 \end{bmatrix}$$

Now

$$D = \begin{bmatrix} 3 & 0 \\ 0 & -4 \end{bmatrix}$$

$A = PDP^{-1}$ (factorization as a product of three matrices)

$$A^k = PD^k P^{-1}$$

Now, powers of the matrix A can be calculated as,

$A^4 = PD^4 P^{-1}$

$$A^4 = -\frac{1}{7}\begin{bmatrix} 1 & 1 \\ -1 & -8 \end{bmatrix}\begin{bmatrix} 81 & 0 \\ 0 & 256 \end{bmatrix}\begin{bmatrix} -8 & -1 \\ 1 & 1 \end{bmatrix}$$

$$A^4 = \begin{bmatrix} 56 & -25 \\ 200 & 281 \end{bmatrix}$$

4.4 Solving Systems of Linear Equations

4.4.1 Introduction

It was shown previously that there are different ways of solving a set of linear equations. One method was to make use of determinants, that is, using Cramer's rule. Given two equations in two unknowns

$$ax + by = e$$

$$cx + dy = f$$

This has solutions

$$\frac{x}{\begin{vmatrix} e & b \\ f & d \end{vmatrix}} = \frac{y}{\begin{vmatrix} a & e \\ c & f \end{vmatrix}} = \frac{1}{\begin{vmatrix} a & b \\ c & d \end{vmatrix}}$$

provided that they exist.

The advantage of this method is that only one of the answers needs to be calculated, if that is all that is needed.

Another option is to make use of the *inverse matrix method* (see Section 4.4.3). Writing the set of equations in matrix form as AX = B, the solution, provided it exists, is given by X = A⁻¹B. The advantage of this method is that results can be obtained for lots of inputs by changing the right-hand set of values and using the same inverse matrix A⁻¹ each time.

Each method has its advantages of course, but they also have the disadvantage that they do not give us much information about situations when there is not a unique solution. As engineers, it is very important to be able to tell what happens in extreme cases! Also, there may be a need to know when there might be more than one solution for a set of equations.

4.4.2   Gaussian Elimination Method

The Gaussian elimination method for solving sets of linear gives more information about no solutions or lots of solutions, as well as finding "the solution" when it exists. It's probably easiest to learn this method from a worked example.

Example 4.19

Solve the set of equations:

$$-x + y + 2z = 2$$
$$3x - y + z = 6$$
$$-x + 3y + 4z = 4$$

Step 1: Write the *augmented matrix*:

$$\begin{pmatrix} -1 & 1 & 2 & \vdots & 2 \\ 3 & -1 & 1 & \vdots & 6 \\ -1 & 3 & 4 & \vdots & 4 \end{pmatrix}$$

Step 2: Perform *elementary row operations* on the augmented matrix until it is in echelon form. *Echelon form* means that the first nonzero element in any row lies to the right of the first nonzero element in the row above. (Note: Some other definitions of echelon form exist, including having a one as the first nonzero element in each row. This definition is not used here, although it is easy to divide each row by a suitable number to get it in this form.) The matrix will have the following structure:

$$\begin{pmatrix} a & b & c & \vdots & d \\ 0 & e & f & \vdots & g \\ 0 & 0 & h & \vdots & i \end{pmatrix}$$

where a, b, c, d, e, f, g, h, and i are constants. The basically idea is to aim for a "triangle of zeros" in the bottom left-hand corner of the matrix.

Elementary row operations can be

- Multiplying any row by a number
- Interchanging any two rows
- Adding to any row a multiple of another row

Generally, all this is easier to do than to explain! It is always a good idea to note what has been done at each step.

Let's get two zeros in the first column of the matrix $\begin{pmatrix} -1 & 1 & 2 \vdots 2 \\ 3 & -1 & 1 \vdots 6 \\ -1 & 3 & 4 \vdots 4 \end{pmatrix}$ by

Row 2 = Row 2 + 3 Row 1,
(R2: R2 + 3R1)

and Row 3 = Row 3 − Row 1,
(R3: R3 − R1)

$$\begin{pmatrix} -1 & 1 & 2 \vdots 2 \\ 0 & 2 & 7 \vdots 12 \\ 0 & 2 & 2 \vdots 2 \end{pmatrix}$$

Now let's get a zero in the second column by
Row 3 = Row 3 − Row 2,
(R3: R3 − R2)

$$\begin{pmatrix} -1 & 1 & 2 \vdots 2 \\ 0 & 2 & 7 \vdots 12 \\ 0 & 0 & -5 \vdots -10 \end{pmatrix}$$

Note: When putting a zero in the second column, one must always use row 2 with a row 3 otherwise using row 1 with row 3 will change the zero value already created in the first element of row 3 and so undoing what has already been achieved.

The objective has been achieved, since the nonzero elements start one place to the right as we go down the rows. The augmented matrix is in echelon form.

Step 3: The solution can now be found by *back-substitution*. Imagine the equations returned to their original form. They would now read

$$-x + y + 2z = 2$$
$$2y + 7z = 12$$
$$-5z = -10$$

From the last equation it can be seen that dividing each side by −5 gives $z = 2$. Then, looking at the middle equation and using the value for z gives $2y + 14 = 12$, so $2y = -2$ and $y = -1$. Finally, moving to the first equation and using the two values that are known already gives $-x + (-1) + 2(2) = 2$, from which $x = 1$. Putting all the results together gives the solutions as

$$x = 1, \qquad y = -1, \qquad z = 2.$$

This is an example where there was a *unique solution* to the set of equations.

Example 4.20: If No Solution Exists

Consider the equations

$$3x + 2y + z = 3$$
$$2x + y + z = 0$$
$$6x + 2y + 4z = -4$$

Let's try the method of Gaussian elimination on this set of equations. First, write the augmented matrix:

$$\begin{pmatrix} 3 & 2 & 1 & \vdots & 3 \\ 2 & 1 & 1 & \vdots & 0 \\ 6 & 2 & 4 & \vdots & 6 \end{pmatrix}$$

Let's get two zeros in the first column by
Row 2 = Row 2 − 2/3 Row 1
(R2: R2 − 2/3 R1)

and Row 3 = Row 3 − 2 Row 1
(R3: R3 − 2R1)

$$\begin{pmatrix} 3 & 2 & 1 & \vdots & 3 \\ 0 & -\frac{1}{3} & \frac{1}{3} & \vdots & -2 \\ 0 & -2 & 2 & \vdots & 0 \end{pmatrix}$$

Now let's get a zero in the second column by
Row 3 = Row 3 − 6 Row 2
(R3 = R3 − 6R2)

$$\begin{pmatrix} 3 & 2 & 1 & \vdots & 3 \\ 0 & -\frac{1}{3} & \frac{1}{3} & \vdots & -2 \\ 0 & 0 & 0 & \vdots & 12 \end{pmatrix}$$

It can be seen that the last row contains a *contradiction*. By writing it out as an equation it becomes $0 = 12$? This tells one that the system of equations has *no solution*. It is said that the system is *inconsistent*.

Example 4.21: If Infinitely Many Solutions Exist

Consider the equations

$$3x + 2y + z = 3$$
$$2x + y + z = 0$$
$$-x \quad - z = 3$$

Let's try the method of Gaussian elimination on this set of equations. First, write the augmented matrix:

$$\begin{pmatrix} 3 & 2 & 1 & \vdots & 3 \\ 2 & 1 & 1 & \vdots & 0 \\ -1 & 0 & -1 & \vdots & 3 \end{pmatrix}$$

Let's get two zeros in the first column by
Row 2 = Row 2 – 2/3 Row 1
(R2: R2 – 2/3 R1)

and Row 3 = Row 3 + 1/3 Row 1
(R3: R3 + 1/3 R1)

$$\begin{pmatrix} 3 & 2 & 1 & \vdots & 3 \\ 0 & -\frac{1}{3} & \frac{1}{3} & \vdots & -2 \\ 0 & \frac{2}{3} & -\frac{2}{3} & \vdots & 4 \end{pmatrix}$$

Now let's get a zero in the second column by
Row 3 = Row 3 + 2 Row 2
(R3: R3 + 2R2)

$$\begin{pmatrix} 3 & 2 & 1 & \vdots & 3 \\ 0 & -\frac{1}{3} & \frac{1}{3} & \vdots & -2 \\ 0 & 0 & 0 & \vdots & 0 \end{pmatrix}$$

The last row gives nothing useful, since 0 = 0. The second row gives

$$-\frac{1}{3}y + \frac{1}{3}z = -2$$

or, multiplying each side by 3,

$$-y + z = -6$$

z is *arbitrary*; in other words, it can take *any value*. Suppose, letting a parameter such as t represent any number for z, that is, $z = t$. Then y can be found from $y = z + 6 = t + 6$. Then, from the first equation, $3x + 2y + z = 3$, gives $3x + 2(t + 6) + t = 3$ or $3x = -3t - 9$. So $x = -t - 3$.
So the "solution" can be given as, $x = -t - 3$, $y = t + 6$, $z = t$.

Any set of numbers that satisfied these equations would also satisfy the original set of simultaneous equations. The parameter t can be replaced with *any real number*. Hence, there are *infinitely* many solutions to this system.

The three situations be generalized as (1) a unique solution, (2) no real solutions, and (3) infinitely many solutions. It's definitely got something to do with the last row (or rows) of the echelon form matrix being zero.

The rank of a matrix is the number of nonzero rows when it has been reduced to echelon form. The three preceding situations finished up looking like the following:

Unique solution: $\begin{pmatrix} & & & \vdots & \\ 0 & 0 & c & \vdots & \end{pmatrix}$ rank(A) = rank(A,d) = 3

       non-zero

Infinite number of solutions: $\begin{pmatrix} & & & \vdots & \\ 0 & 0 & 0 & \vdots & 0 \end{pmatrix}$ rank(A) = rank(A,d) = 2

No solutions: $\begin{pmatrix} & & & \vdots & \\ 0 & 0 & 0 & \vdots & c \end{pmatrix}$ rank(A) = 2, rank(A,d) = 3

       non-zero

Similar rules can be applied to any number of equations.

4.4.3 Matrix Inversion Method

Two simultaneous linear equations can be written out in matrix form. Consider the following simultaneous equations:

$$x + 2y = 4$$
$$3x - 5y = 1$$

These can be written in matrix form as

$$\begin{pmatrix} 1 & 2 \\ 3 & -5 \end{pmatrix} \begin{pmatrix} x \\ y \end{pmatrix} = \begin{pmatrix} 4 \\ 1 \end{pmatrix}$$

Denoting

$$A = \begin{pmatrix} 1 & 2 \\ 3 & -5 \end{pmatrix}, \quad X = \begin{pmatrix} x \\ y \end{pmatrix}, \quad B = \begin{pmatrix} 4 \\ 1 \end{pmatrix}$$

This gives the following matrix equation $AX = B$. This is the matrix of the simultaneous equations. Here the unknown is the matrix X, since A and B are already known. A is called the *matrix of coefficients.*

4.4.3.1 Matrix Method for Solving Simultaneous Equations

Given the system of equations in matrix form

$$AX = B \tag{4.12}$$

Multiplying both sides of Equation 4.12 by the inverse of A, provided it exists, gives

$$A^{-1}AX = A^{-1}B \tag{4.13}$$

But $A^{-1}A = I$, the identity matrix. Also, $IX = X$, so this leaves

$$X = A^{-1}B \tag{4.14}$$

This result given by Equation 4.14 gives a method for solving simultaneous equations. First write the equations in matrix form, calculate the inverse of the matrix of coefficients A^{-1}, and finally perform a matrix multiplication.

Note: When $det(A) = 0$ there are no solutions and so the matrix is not invertible because of division-by-zero problems.

Example 4.22

Solve the simultaneous equations

$$x + 2y = 4$$
$$3x - 5y = 1$$

Solution: Putting these in matrix form gives

$$\begin{pmatrix} 1 & 2 \\ 3 & -5 \end{pmatrix} \begin{pmatrix} x \\ y \end{pmatrix} = \begin{pmatrix} 4 \\ 1 \end{pmatrix}$$

$$AX = B$$

Now calculate the inverse of $A = \begin{pmatrix} 1 & 2 \\ 3 & -5 \end{pmatrix}$:

$$A^{-1} = -\frac{1}{11} \begin{pmatrix} -5 & -2 \\ -3 & 1 \end{pmatrix}$$

Then X is given by using

$$X = A^{-1} B = -\frac{1}{11} \begin{pmatrix} -5 & -2 \\ -3 & 1 \end{pmatrix} \begin{pmatrix} 4 \\ 1 \end{pmatrix}$$

$$= -\frac{1}{11} \begin{pmatrix} -22 \\ -11 \end{pmatrix}$$

$$= \begin{pmatrix} 2 \\ 1 \end{pmatrix}$$

Hence $x = 2$, $y = 1$ are the solutions.

The same method can now be used to solve a system of three simultaneous equations as shown in the next example.

Example 4.23

Solve the simultaneous equations

$$-x - 2y + 2z = 1$$
$$2x + y + z = 7$$
$$3x + 4y + 5z = 26$$

Solution: These can be put into a matrix equation as

$$\begin{pmatrix} -1 & -2 & 2 \\ 2 & 1 & 1 \\ 3 & 4 & 5 \end{pmatrix} \begin{pmatrix} x \\ y \\ z \end{pmatrix} = \begin{pmatrix} 1 \\ 7 \\ 26 \end{pmatrix}$$

$$A\,X = B$$

The inverse of A is calculated as

$$A^{-1} = \frac{1}{23} \begin{pmatrix} 1 & 18 & -4 \\ -7 & -11 & 5 \\ 5 & -2 & 3 \end{pmatrix}$$

To find the solution

$$X = A^{-1}B = \frac{1}{23} \begin{pmatrix} 1 & 18 & -4 \\ -7 & -11 & 5 \\ 5 & -2 & 3 \end{pmatrix} \begin{pmatrix} 1 \\ 7 \\ 26 \end{pmatrix} = \frac{1}{23} \begin{pmatrix} 23 \\ 46 \\ 69 \end{pmatrix}$$

$$X = \begin{pmatrix} 1 \\ 2 \\ 3 \end{pmatrix}$$

that is, $x = 1$, $y = 2$, and $z = 3$.

4.5 Applications

Example 4.24: Fire Modeling Using Markov Chains

A Markov chain is a process that satisfies the Markov property, meaning one can make predictions for the future of the process based solely on its present state just as well as one could knowing the process's full history. The process involves the probabilities of being in states and can be modeled using matrices. The use of matrices and matrix operations can provide solutions to problems and is useful to consider them here.

Consider the four different states of a fire as follows: O, fire is out; S, smoke development; F, flashover; and B, full burning fire. The time steps are in minutes and Figure 4.1 shows the transition diagram modeling the fire with transition probabilities between the different states. It is required to find how long it will take the fire in the different states to go out, that is, to reach the absorbing state O.

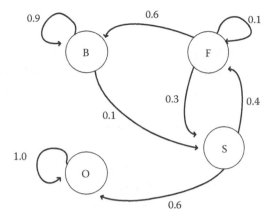

Figure 4.1 Transition diagram modeling a fire.

For this setup, a transition matrix P can be derived from the transition diagram for the different states as follows,

$$P = \begin{array}{c} \\ O \\ S \\ F \\ B \end{array} \begin{array}{cccc} O & S & F & B \\ \hline \left(\begin{array}{c|ccc} 1 & 0 & 0 & 0 \\ \hline 0.6 & 0 & 0.4 & 0 \\ 0 & 0.3 & 0.1 & 0.6 \\ 0 & 0.1 & 0 & 0.9 \end{array} \right) \end{array}$$

Within this transition matrix P, there is a smaller submatrix that can be formed by removing the absorbing state O and is defined as Q, where

$$Q = \begin{pmatrix} 0 & 0.4 & 0 \\ 0.3 & 0.1 & 0.6 \\ 0.1 & 0 & 0.9 \end{pmatrix}$$

It turns out from the theory that an important fundamental matrix F_M gives the times from the different states to the absorbing state and can be formed using

$$F_M = (I - Q)^{-1}$$

where identity matrix $I = I_3$ in this case and the power -1 indicates the inverse of the $(I - Q)$ matrix.

Constructing

$$(I-Q) = \begin{pmatrix} 1 & 0 & 0 \\ 0 & 1 & 0 \\ 0 & 0 & 1 \end{pmatrix} - \begin{pmatrix} 0 & 0.4 & 0 \\ 0.3 & 0.1 & 0.6 \\ 0.1 & 0 & 0.9 \end{pmatrix} = \begin{pmatrix} 1 & -0.4 & 0 \\ -0.3 & 0.9 & -0.6 \\ -0.1 & 0 & 0.1 \end{pmatrix}$$

and calculating the inverse of this 3×3 matrix gives

$$F_M = (I-Q)^{-1} = \begin{array}{c} S \\ F \\ B \end{array} \begin{pmatrix} 1.67 & 0.74 & 4.4 \\ 1.67 & 1.85 & 11.1 \\ 1.67 & 0.74 & 14.4 \end{pmatrix}$$

So, to obtain the average time to get from the different fire states to the absorbing state (fire out) is given by adding along each row as follows:

Smoking state, S: $1.67 + 0.74 + 4.4 = 6.8$ minutes
Flashover state, F: $1.67 + 1.85 + 11.1 = 14.6$ minutes
Full burn state, B: $1.67 + 0.74 + 14.4 = 16.8$ minutes

Example 4.25: Diagonal Factorization in Difference Equations

Given a first-order difference equation $u_{k+1} = Au_k$ governing some system behavior with given start vector u_0, this equation can be solved as follows:

$$u_1 = Au_0$$
$$u_2 = Au_1 = A^2 u_0$$
$$u_3 = Au_2 = AA^2 u_0 = A^3 u_0$$
$$\vdots$$

$u_k = A^k u_0$ is the general solution to the equation for any k.

Now, to find what will be the state after, say, 100 iteration or time intervals, letting $k = 100$ gives

$$u_{100} = A^{100} u_0$$

To calculate A^{100} using just matrix multiplication is very difficult, but if one can diagonalize A using its eigenvalues and eigenvectors as $A = PDP^{-1}$, then it is much easier to find A^{100}.

Note: This type of difference equation $u_{k+1} = Au_k$ is seen to appear in many practical applications such as Markov Chains in the area of probabilistic risk. Also, u_k could be used to represent the position in an evacuation model on a floor grid during a fire.

Example 4.26:   Dimensional Analysis in Fluid Mechanics

Consider the problem of how the drag force is affected on a smooth sphere in some uniform fluid flow. The important parameters that can affect the drag force F might be D the diameter of the sphere, v the velocity of the fluid, ρ the density of fluid, and μ the viscosity of the fluid. This can be represented as $F = f(D, v, \rho, \mu)$, that is, F is some function of these four parameters that needs to be determined. If experiments were to be carried out to see this dependence, say, 10 values of varying the diameter D while fixing the other three parameters could be carried out. In the same way, varying the other three parameters would have to be conducted as well. This will require a lot of experiments and the time taken to do all these would be very large and in reality impractical to do.

What is needed is a more efficient method and dimensional analysis is a way that can reduce the problem to a more realistic situation in which only a few experiments need to be carried out. The method used to solve this particular drag force problem uses the Buckingham Pi theorem, which is not discussed here. However, within this method the property of *dimensional homogeneity* is used. *All equations will have the same units on both sides of the defining equation* and it is this property that then requires solving a set of linear simultaneous equations.

To illustrate this idea of dimensional homogeneity further consider a simpler problem. In mechanics, the horizontal range R traveled by a projectile can depend on the parameters horizontal velocity V_x, the vertical velocity V_y, and the gravitational acceleration g, that is, can say that $R = f(V_x, V_y, g)$. To determine this actual relationship, the units of the parameters are first determined with velocity (meters/second) and acceleration (meters/seconds squared). Using L = meter, T = time, L_x = length in horizontal direction, and L_y = length in the vertical direction, then the range R can be represented as

$$R \propto V_x^a V_y^b g^c \tag{4.15}$$

where the parameters a, b, and c are powers to be determined using dimensional homogeneity. Putting in the units for both sides of Equation 4.15 gives

$$\left[L_x\right]^1 = \left[L_x \, T^{-1}\right]^a \left[L_y \, T^{-1}\right]^b \left[L_y \, T^{-2}\right]^c$$

Considering the different dimensions and balancing the powers on both sides gives the following:

$$L_x : 1 = a, \qquad L_y : 0 = b + c, \qquad T : 0 = -a - b - 2c$$

Now these form a system of three equations in three unknowns that need to be solved to find a, b, and c, and hence give the form of relationship for R. These can be solved by different methods but here the Gaussian elimination method of Section 4.4.2 is used as follows:

$$a + b + 2c = 0$$
$$b + c = 0$$
$$a = 1$$

First, the augmented matrix:

$$\begin{pmatrix} 1 & 1 & 2 & \vdots & 0 \\ 0 & 1 & 1 & \vdots & 0 \\ 1 & 0 & 0 & \vdots & 1 \end{pmatrix}$$

R3: R3 − R1 gives

$$\begin{pmatrix} 1 & 1 & 2 & \vdots & 0 \\ 0 & 1 & 1 & \vdots & 0 \\ 0 & -1 & -2 & \vdots & 1 \end{pmatrix}$$

R3: R3 + R2 gives

$$\begin{pmatrix} 1 & 1 & 2 & \vdots & 0 \\ 0 & 1 & 1 & \vdots & 0 \\ 0 & 0 & -1 & \vdots & 1 \end{pmatrix}$$

The equations now read as,

$$a + b + 2c = 0$$
$$b + c = 0$$
$$-c = 1$$

From this it is seen that $c = -1$, $b = 1$, and $a = 1$, taking these values and substituting them back into Equation 4.15 gives the relationship for the range as

$$R \propto \frac{V_x V_y}{g}$$

Problems

4.1 Calculate the following determinants.

a. $\begin{vmatrix} 5 & 1 \\ 3 & -2 \end{vmatrix}$

b. $\begin{vmatrix} 4 & 1 & 3 \\ -1 & 2 & 7 \\ 1 & 5 & 0 \end{vmatrix}$

4.2 Find the value of the constant k for which $\begin{vmatrix} 1 & k & 2 \\ k-1 & 0 & 1 \\ 1 & 2 & 0 \end{vmatrix} = 0$

4.3 Let $A = \begin{pmatrix} 2 & 1 & 5 \\ 1 & 3 & -2 \end{pmatrix}$ and $B = \begin{pmatrix} 4 & 2 & 1 \\ 3 & -1 & 7 \end{pmatrix}$, calculate the following.

   a. $A + B$

   b. $A - B$

4.4 Given

$$A = \begin{bmatrix} 4 & 1 \\ 3 & 2 \\ 5 & 0 \end{bmatrix} \quad \text{and} \quad B = \begin{bmatrix} 2 & -3 \\ 1 & 4 \end{bmatrix}$$

   calculate the following.

   a. AB b. B^3 c. AA^T

4.5 Solve the following set of simultaneous equations using the Gaussian elimination method.

   a. $x + 2y + z = 7$ b. $x + y - z = 2$
   $3x + y + 4z = 5$ $x + 2y = 5$
   $2x + 3y - z = 14$ $2x + 3y - z = 7$

4.6 An electrical network has currents i_1, i_2, and i_3 given by the following set of equations:

$$i_1 + 3i_2 + i_3 = 9$$
$$2i_1 - i_2 + 5i_3 = 7$$
$$4i_1 + 2i_2 - i_3 = 11$$

Using the matrix inversion method, find the currents in the network.

4.7 For the matrix

$$A = \begin{pmatrix} 2 & 1 & 3 \\ 1 & 2 & 3 \\ 3 & 3 & 20 \end{pmatrix}$$

   a. Show that the characteristic equation given by det $(\lambda I - A) = 0$ is $\lambda^3 - 24\lambda^2 + 65\lambda - 42 = 0$.

b. Hence, find the eigenvalues and eigenvectors of the matrix A.

4.8 Given the 2×2 matrix $A = \begin{pmatrix} 2 & 1 \\ 9 & 2 \end{pmatrix}$, determine the following.

a. Diagonal Factorization of the matrix A.

b. Hence, calculate the matrix A^{10}.

4.9 Consider again Example 4.24 with the four states of fire: O, fire is out; S, smoke development; F, flashover; and B, full burning fire. The time steps are given in minutes and Figure 4.2 shows the transition diagram modeling the fire with transition probabilities between the different states. Calculate how long it will take the fire in the different states to reach the absorbing state O (i.e., to go out).

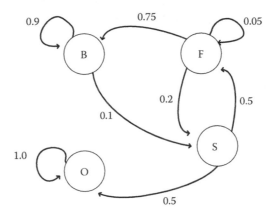

Figure 4.2 Transition diagram modeling a fire with the probabilities between states.

5 Complex Numbers

5.1 Background

The origins of complex numbers came about when mathematicians were trying to solve certain types of equations. Equations of the form $x^2 = 9$, which had two real solutions $x = 3$ and $x = -3$, were of no problem. But when the equation was of the form $x^2 = -9$, this has no real solutions since the square of a real number cannot be negative. This was a problem. The idea to overcome this was to extend the real numbers with the imaginary unit called $j = \sqrt{-1}$, where $j^2 = -1$, so that solutions to equations such as $x^2 = -9$ could now be found.

Originally, the symbol i was used to represent $\sqrt{-1}$, but in many engineering fields i was the symbol for electrical current and so the symbol j is used in engineering instead.

5.2 Introduction and the Imaginary j

When solving a quadratic equation using the formula such as the general quadratic equation $ax^2 + bx + c = 0$, where a, b, and c are numbers. Then the solutions are given by the formula

$$x = \frac{-b \pm \sqrt{b^2 - 4ac}}{2a} \tag{5.1}$$

Notes:

- If the number under the square root sign is positive, there are two real and distinct solutions.

- If the number under the square root sign is zero, there is one repeated solution.

- If the number under the square root sign is negative, there are no real solutions.

In the next section, it will be seen that in the third case the solutions can in fact be written as *complex numbers*.

Now, what if the quadratic equation is given as $x^2 + 6x + 10 = 0$? Using the formula given by Equation 5.1 with $a = 1$, $b = 6$, and $c = 10$ gives

$$x = \frac{-6 \pm \sqrt{36 - 40}}{2} = \frac{-6 \pm \sqrt{-4}}{2}$$

Unfortunately, the calculator cannot work out $\sqrt{-4}$. In fact, with "normal" (or real) numbers, there is not any number that when multiplied by itself can give a negative result.

But suppose there was an invented "number," call it j, such that $j^2 = -1$. What then? It can be seen that

$$(2j)(2j) = 4j^2 = 4(-1) = -4$$

and

$$(-2j)(-2j) = 4j^2 = 4(-1) = -4$$

So $\sqrt{-4}$ can be written as $2j$ or $-2j$ and the solutions to the quadratic equation can now be written as

$$x = \frac{-6 \pm 2j}{2} = -3 + j \quad \text{or} \quad -3 - j$$

These are *complex solutions* to the quadratic equation.

5.2.1   Some Properties of j

If $j^2 = -1$, then $j^3 = -j$, $j^4 = 1$, $j^5 = j$, and so on. You could also write $\sqrt{-4} = \pm 2j$, $\sqrt{-9} = \pm 3j$, $\sqrt{-16} = \pm 4j$, and so on.

5.2.2   Complex numbers:

A complex number is a number like $2 + 3j$. The *real part* of $2 + 3j$ is 2. The *imaginary part* of $2 + 3j$ is 3. (Note: The imaginary part doesn't include j.)

In any equation with complex numbers, the real part of one side has to equal the real part of the other side; similarly for the imaginary parts. There are two equations rolled up as one. If it is given that $x + yj = 3 - 4j$, then it can be seen straight away that $x = 3$ and $y = -4$.

5.3   Arithmetic Operations

5.3.1   Addition and Subtraction

It is easy to add and subtract complex numbers. It goes just like ordinary algebra; add the real parts together and add the imaginary parts together.

Example 5.1

1. $(3 + j) + (1 + 2j) = 4 + 3j$
2. $(2 - 3j) - (1 + j) = 1 - 4j$
3. $(1 + 4j) + 2(5 - j) = 11 + 2j$

5.3.2 Multiplication

Multiplying complex numbers is also simple, as long as you remember that $j^2 = -1$.

Example 5.2

1. $j(3 + j) = 3j + j^2 = -1 + 3j$
2. $(2 + j)(4 + 5j) = 2(4 + 5j) + j(4 + 5j) = 8 + 10j + 4j + 5j^2 = 3 + 14j$

5.3.2.1 Conjugate Numbers

When multiplying complex pairs such as $(2 - j)(2 + j) = 5$ and $(1 + 6j)(1 - 6j) = 37$ notice here that each time that the result is a real number. The "partner" of the complex number is called its *conjugate*. So the conjugate of $2 - j$ is $2 + j$. The conjugate of $2 + j$ is $2 - j$. The conjugate of $-5 - 7j$ is $-5 + 7j$. The conjugate of $100\,j$ is $-100\,j$. The conjugate of 27 is 27. The conjugate of $a + bj$ is $a - bj$. Also notice that the result of multiplying a number by its conjugate is always real and positive. Generally, the following is found:

$$(a + bj)(a - bj) = a^2 + b^2$$

The conjugate of the complex number z is written either as $z*$ or as $\bar{z}$. So if $z = 2 + 3j$, then $z* = 2 - 3j$ and $zz* = 13$. And if $z = 1 + 5j$, then $z* = 1 - 5j$, and $zz* = 26$.

5.3.3 Division

When it comes to division of complex numbers, it makes no sense to be dividing by imaginary numbers. This idea of the complex conjugate helps out here as an equivalent division can be done such that the denominator is now a real number as seen in the next example.

Example 5.3

$$\frac{1}{2+j} = \frac{1(2-j)}{(2+j)(2-j)} = \frac{2-j}{2^2+1^2} = \frac{2-j}{5} = \frac{2}{5} - \frac{1}{5}j = 0.4 - 0.2j$$

Here, the conjugate of the denominator $(2 + j)$ is $(2 - j)$, and this is multiplied top and bottom such that the denominator becomes a real number, which is 5 in this case.

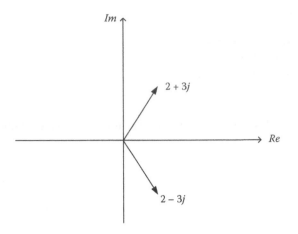

Figure 5.1 Argand diagram showing complex number representation.

5.4 Argand Diagram

5.4.1 Drawing a Diagram of Complex Numbers

It is often helpful to draw the *complex plane*, or *Argand diagram*, and represent the complex number $2 + 3j$, say, as a vector starting at the origin and finishing at the point (2,3) as shown in Figure 5.1.

Note: Adding and subtracting complex numbers follows the rules of vector addition and subtraction.

5.5 Polar and Exponential Form

5.5.1 Polar Form

Figure 5.2 shows the diagram of the complex plane showing the number $4 + 3j$. The complex number $4 + 3j$ can also be written as a distance from the origin to the point and an angle measured anticlockwise from the real axis as (5 cos 36.9°) + j(5 sin 36.9°) or 5(cos 36.9° + j sin 36.9°). This is called the *polar form* of the number. ($4 + 3j$ is called the *rectangular form*). Sometimes the polar form

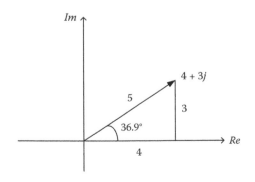

Figure 5.2 Different representations of the complex number.

is written in shorthand to $5 \angle 36.9°$. (The length is called the modulus. The angle is called the argument.)

Note: The angle can be expressed in several ways, for example, $-30°$ could be written as $330°$; $200°$ could be written as $-160°$.

Generally, a complex number in rectangular form is $a + bj$, as shown in Figure 5.3.

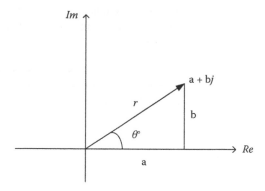

Figure 5.3 General point $a + bj$ in the Argand diagram.

It can be seen that the modulus r is given by the Pythagorean theorem as $r = \sqrt{a^2 + b^2}$ and the angle θ is obtained using basic trigonometry and in this case as $\theta = \tan^{-1} \dfrac{b}{a}$.

An important property of the polar form is that this angle θ is not unique and it is the case that $\theta + 360° \, n$ (where $n = 0, 1, 2,...$) would also be at the same position in the complex plane. This will be very important when calculating roots of equations in Section 5.6.

Also, the following relationship exists to give the polar form as

$$z = a + bj = r\cos\theta + r\sin\theta j = r(\cos\theta + j\sin\theta)$$

Note: Most calculators can easily swap a number from one form of a complex number to the other using the Pol and Rec function buttons.

5.5.1.1 Multiplying and Dividing Complex Numbers in Polar Form

The polar form makes it very easy to multiply or divide complex numbers. Because of the way sines and cosines work

$$(5 \angle 25°)(2 \angle 60°) = (5 \times 2) \angle (25° + 60°) = 10 \angle 85°$$

Or when multiplying two numbers in polar form, just multiply their lengths and add their angles.

Similarly,

$$\frac{5\angle25°}{6\angle60°} = \left(\frac{5}{2}\right)\angle(25° - 60°) = 2.5\angle(-35°)$$

Or when dividing two numbers in polar form, just divide their lengths and subtract their angles.

5.5.2 Exponential Form

Using Euler's identity, $e^{j\theta} = \cos\theta + j\sin\theta$. Then, $z = r(\cos\theta + j\sin\theta) = r\,e^{j\theta}$. In this form r is the modulus and θ is the argument in radians. This is known as the *exponential form* of a complex number.

Note: The proof of Euler's identity uses the Taylor series expansions of the exponential, cosine, and sine functions, and is omitted here.

5.5.3 Powers of Complex Numbers

Multiplying a complex number by itself gives a power of that number:

$$(5\angle25°)(5\angle25°)(5\angle25°)(5\angle25°) = (5\angle25°)^4$$

Because of the way multiplication works with complex numbers in polar form, this can be worked out easily as

$$(5\times5\times5\times5)\angle(25° + 25° + 25° + 25°) = 5^4\angle(4\times25°).$$

This shows an important result that

$$(5\angle25°)^4 \equiv 5^4\angle(4\times25°)$$

This result is known as De Moivre's theorem and is stated more generally in the next section.

5.5.4 De Moivre's Theorem

De Moivre's theorem relates to raising powers of complex numbers in polar form as follows:

$$\{r(\cos\theta + j\sin\theta)\}^n = r^n(\cos n\theta + j\sin n\theta) \qquad (5.2a)$$

Or

$$\{r\angle\theta\}^n = r^n\angle n\theta \qquad \text{(in shortened form)} \qquad (5.2b)$$

The result works for both positive and negative whole number powers and is used for finding the roots of equations in the next section.

5.6   Roots of Equations

De Moivre's theorem can also be used with fractional powers to find square roots, cube roots, and so forth of complex numbers. There's a difference, however: This time, there should be more than one answer. To see why, try calculating z^3 if $z = 2 \angle 60°$.

From De Moivre,

$$z^3 = \{2 \angle 60°\}^3 = 8 \angle 180°$$

and this can be expressed in rectangular form as –8. If it is drawn on the complex plane, it will look like Figure 5.4.

But suppose one tries the same thing with the complex number $w = 2 \angle 300°$. Then $z^3 = 8 \angle 900°$, and turning this answer into rectangular form, it's the same as before: –8. Are there any other numbers that would give the answer –8 when cubed? Without knowing anything about complex numbers, one can clearly say that $(-2)^3 = -8$. Therefore, –2 can be written as a complex number in polar form as $2 \angle 180°$.

So there appears to be three different complex numbers that, when cubed, give the answer –8. These are as follows: $z_1 = 2 \angle 60°$, $z_2 = 2 \angle 180°$, $z_3 = 2 \angle 300°$. These are the three cube roots of –8, or the three values of $(-8)^{\frac{1}{3}}$. If these are drawn on the complex plane, they will look like Figure 5.5. Notice, how they are equally spaced (120°) around a circle.

The procedure to find all the different values for the roots is very straightforward. Suppose one needs to find $z^{\frac{1}{4}}$, if z is the complex number $81 \angle 160°$. This can be done by completing the following steps:

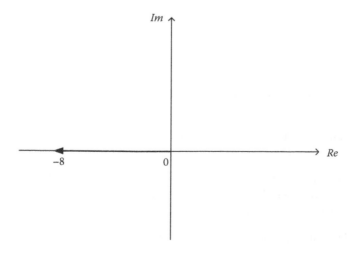

Figure 5.4 Number –8 on the Argand diagram.

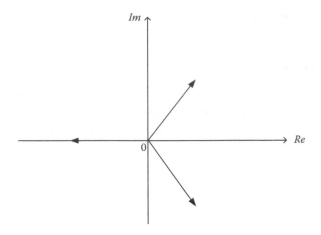

Figure 5.5 The three cube roots of −8 on the Argand diagram.

Step 1: $z^{\frac{1}{4}} = \{81\angle160°\}^{\frac{1}{4}}$ Make sure the numbers are in polar form.

Step 2: $= \{81\angle(160° + 360° \; n)\}^{\frac{1}{4}}$ Add $360°n$ to the angle.

Step 3: $= 81^{\frac{1}{4}}\angle\frac{1}{4}(160° + 360°n)$ Using De Moivre.

$= 3\angle(40° + 90°n)$ Tidying up.

Step 4: $= \begin{cases} 3\angle40° \\ 3\angle130° \\ 3\angle220° \\ 3\angle310° \end{cases}$ Taking the values of $n = 0, 1, 2, 3…$

To draw these, they would be equally spaced (90°) around a circle.

Notes:

- Starting with a number that is not in polar form, the first task is to turn it into polar form.

- The answers are arrived at by taking $n = 0, 1, 2, 3, …$ and go on until they start to repeat. In the earlier example, the next one would be $3\angle400°$, which is the same as $3\angle40°$, which is the first solution again.

- The number of solutions to expect is given by looking at the bottom number of the fractional power; for example, $z^{\frac{1}{5}}$ will have five values equally spaced around a circle and $z^{\frac{1}{7}}$ will have seven values equally spaced around a circle. So, for the mth root there are m values.

- Having found the set of answers in polar form, these can easily be turned into rectangular form if required.

- The *principal root* is defined as the one closest to the positive real axis.

Example 5.4

Determine all solutions of the equation $z^3 - 125\,j = 0$, giving your answers in rectangular form.

Solution:

$z^3 - 125\,j = 0$

$\therefore z^3 = 125\,j$ Rearranging into standard equation form.

$\therefore z = (125\,j)^{\frac{1}{3}}$ Taking both sides to the power $\frac{1}{3}$.

$= \{125 \angle 90°\}^{\frac{1}{3}}$ Change the number into polar form.

$= \{125 \angle (90° + 360°\,n)\}^{\frac{1}{3}}$ Add $360°\,n$ to the angle.

$= 125^{\frac{1}{3}} \angle \frac{1}{3}(90° + 360°\,n)$ Using De Moivre's theorem.

$= 5 \angle (30° + 120°\,n)$ Tidying up.

$$= \begin{cases} 5 \angle 30° \\ 5 \angle 150° \\ 5 \angle 270° \end{cases}$$ Taking enough values of $n = 0, 1, 2$, to get all solutions.

$$= \begin{cases} 4.33 + 2.5\,j \\ -4.33 + 2.5\,j \\ -5\,j \end{cases}$$ Change into rectangular form if required.

5.7   Applications

The use of complex numbers occurs in many areas of science and engineering, and one such application is in determining conditions for flashover fires and disasters to occur. Another important area is in electrical circuit theory with alternating currents, resistor, inductor, and capacitor circuits.

Example 5.5: Conditions for Flashover Fires to Occur

In a room on fire with a single door and a single fire bed, fire modeling can lead to a simplified equation of the temperature T against time t:

$$K\frac{dT}{dt} = a_2 T^2 - a_1 T + 1 \tag{5.3}$$

where K is a constant depending on mass and specific heat capacity. The right-hand side of Equation 5.3 is a quadratic equation in T with defining parameters a_1 and a_2. From theory it turns out that if the roots of this quadratic equation are complex in nature, then a flashover fire will occur.

As an example if $a_1 = 0.01$ and $a_2 = 0.2$ then the quadratic equation on the right-hand side (RHS) becomes

$$0.2T^2 - 0.01T + 1 = 0$$

Solving this using Equation 5.1 gives

$$T = \frac{0.01 \pm \sqrt{-0.7999}}{0.4}$$

which gives the two solutions as $T = 0.025 \pm j\,2.236$, which are two complex conjugate roots. Hence for these parameter values the fire would develop into a flashover state and have possible disastrous consequences.

Example 5.6: Representing Phases on the Complex Plane

Consider the RCL electrical circuit in Figure 5.6. From some basic electrical circuit theory, the resistor has effective resistance R (ohms), the inductor L (henrys) has effective resistance (known as reactance) X_L, and the capacitor C (farads) has effective resistance (known as reactance) X_C.

From experiment, these are found to be as follows:

$$X_C = \frac{1}{wC}$$

$$X_L = wL$$

where $w = 2\pi f$ and f is the frequency of the source.

Now, to consider the total resistance of the series circuit one needs to consider all three components, that is, R, L, and C. Again, from experiment

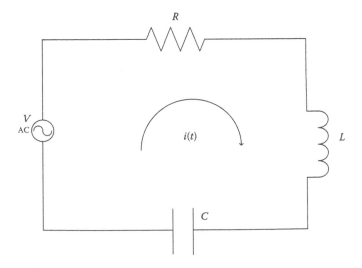

Figure 5.6 An electrical RCL circuit.

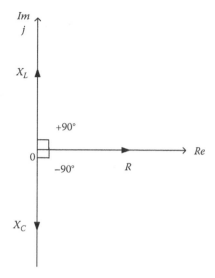

Figure 5.7 The different phases for the component on the complex plane.

it is found that the current through the different components is either in phase or out of phase with the voltage source V.

A phase diagram for the three components showing these phase shifts can be drawn. For the resistor there is no phase shift of the current to the voltage. For the inductor the voltage leads the current by $+90°$. For the capacitor the voltage lags the current by $-90°$. To keep the resistances and reactances separate due to the different phases, these can be put onto a diagram using the real and imaginary axis for the corresponding phases. The phasor diagram is shown in Figure 5.7.

So, knowing the values of R, L, C, and f for a given circuit, then the resistances and reactances for the components can be calculated.

Example 5.7: Impedances and Currents in Electrical Circuits

Consider the circuit shown in Figure 5.8. To calculate the total impedance for the circuit usually, called Z, add all the real and imaginary components together as follows:

$$Z = 4 + j10 + 3 + (-j6)$$

$$Z = 7 + j4$$

Now to find the magnitude and phase angle of the total impedance Z, this complex impedance can be converted into polar form to give $Z = 8.06 \angle 29.7°$, which gives $|Z| = 8.06 \, \Omega$.

Note: Other quantities like the current in the circuit can be found using complex number division, that is, using the relationship $= \dfrac{V}{Z}$.

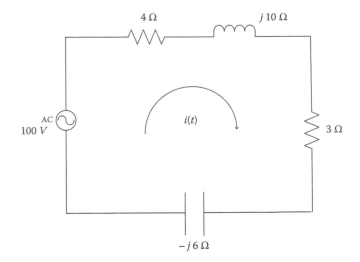

Figure 5.8 An electrical series circuit with resistors, inductor and capacitor.

Example 5.8: Application of complex Numbers to Forces

Consider the force diagram in Figure 5.9. To find the resultant force and the angle at which the force is acting, the forces can first be represented as complex numbers in polar form and then in rectangular form.

The resultant force is given by $F_R = F_1 + F_2 + F_3$. Using the polar form of a complex number as $z = r \cos\theta + r \sin\theta\, j$, each of the three forces can be converted first into polar form then rectangular form as

$$F_1 = 50\cos 45 + 50\sin 45\, j = 35.36 + 35.36\, j$$

$$F_2 = 60\cos 150 + 60\sin 150\, j = -51.96 + 30\, j$$

$$F_3 = 40\cos 255 + 40\sin 255\, j = -10.35 - 38.64\, j$$

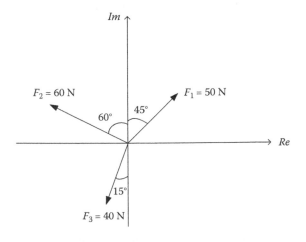

Figure 5.9 Forces acting on a body.

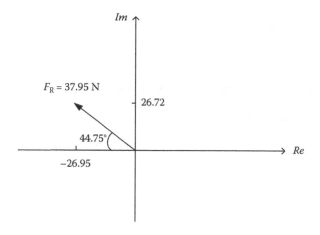

Figure 5.10 Resultant force F_R on the Argand diagram.

So, the result force is $F_R = (35.36 + 35.36\,j) + (-51.96 + 30\,j) + (-10.35 - 38.64\,j)$. Now, just add all the complex numbers to give, $F_R = -26.95 + 26.72\,j$.

This can be put onto an Argand diagram as shown in Figure 5.10. So, the resultant force has a magnitude $|F_R| = 37.95\,\text{N}$ and acts at an angle of $135.25°$ in an anticlockwise direction from the real axis.

Problems

5.1 Solve the following quadratic equations:

    a. $x^2 + 4x + 5 = 0$

    b. $2x^2 - x + 7 = 0$

    c. $3x^2 + 6x + 11 = 0$

5.2 Given that $z_1 = 3 + 5j$ and $z_2 = 1 + 2j$, find the following:

    a. $z_1 + z_2$

    b. $z_1 - 3z_2$

    c. $z_1 z_2$

    d. $\dfrac{z_1}{z_2}$

5.3 Solve the following equation to find the two pairs of values of a and b:
$(a + 3j)(4 - bj) = 37 + 9j$.

5.4 Find the roots of the following equations, giving your answers in rectangular form.

    a. $z^3 = 5 + 4j$

    b. $z^5 - 2 + \sqrt{3}\,j = 0$

c. $z^2 = 18\sqrt{3} - 18j$

d. $z^3 + 27j = 0$

5.5 Consider the electrical circuit in Figure 5.11.

   a. Find the total impedance Z for the circuit.

   b. Hence, find the size of the current i in the circuit, giving the answer in milliamps.

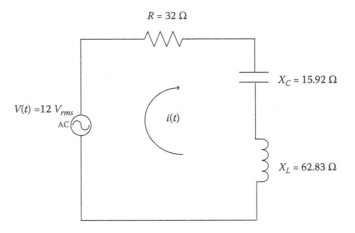

$R = 32\ \Omega$

$X_C = 15.92\ \Omega$

$V(t) = 12\ V_{rms}$
AC

$i(t)$

$X_L = 62.83\ \Omega$

Figure 5.11 An electrical *RCL* series circuit.

5.6 For a room fire the equation for the temperature T against time t is given by Equation 5.3 as

$$K\frac{dT}{dt} = a_2 T^2 - a_1 T + 1.$$

Determine the relationship between the parameters a_1 and a_2 for flash-over to occur.

6 Introduction to Calculus

6.1 Differentiation

6.1.1 Definition of a Limit

For a straight line, to find the *slope* (or *gradient*) of the line, which is the same across the whole line, it is the change in y against the change in x (Figure 6.1).

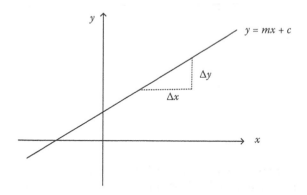

Figure 6.1 The slope of a line given as a change in y against a change in x.

The formula for the slope is given as m, where

$$m = \frac{\text{change in } y}{\text{change in } x} = \frac{\Delta y}{\Delta x} \tag{6.1}$$

What happens if we have some general curve as shown in Figure 6.2. What is meant by the slope of this curve? Clearly, the slope is changing at different points.

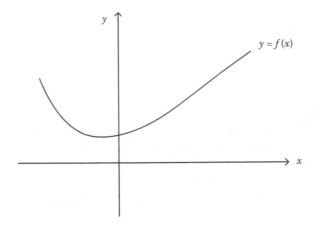

Figure 6.2 Finding slopes of a general curve.

So, the slope at a given point P can be found by saying that this is also the *slope of the tangent line* at that point, as shown in Figure 6.3.

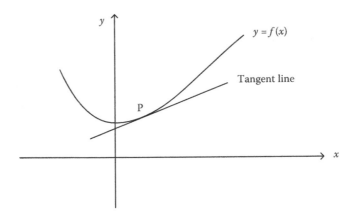

Figure 6.3 The slope of a curve as the slope of the tangent line.

At different points the slope would be different because of the different tangent lines at those points. To find the slope at the different points derivatives and limits are used. Consider the curve given in Figure 6.4. To find the slope at point P, another point Q near to P is considered. The slope of the line joining P to Q, that is, the secant line, is an approximation to the slope at P. So, the closer Q gets to P the better the approximation of the slope at P. The slope of the line PQ is given by using Equation 6.1 as

$$\text{Slope of PQ} = \frac{f(x+h)-f(x)}{x+h - x} = \frac{f(x+h)-f(x)}{h} \tag{6.2}$$

Now by letting $h \to 0$, the line PQ becomes the tangent line at P to the curve, which is the just slope of the curve at the point P. Therefore, the slope at the point P is defined as

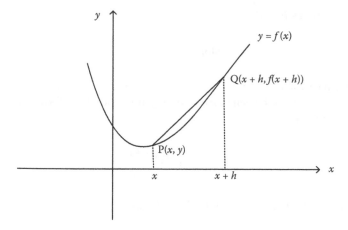

Figure 6.4 Finding the slope of a general curve at a point P.

$$\text{Slope} = \frac{dy}{dx} = \lim_{h \to 0} \frac{f(x+h) - f(x)}{h} \tag{6.3}$$

This is the definition of the derivative, that is, the slope of a curve at a particular point P. This definition can be used to find slopes of different curves and to derive a general formula.

Example 6.1

Review the curve $y = x^2$ in Figure 6.5. Using the formula given by Equation 6.3, the slope of PQ is given as

$$\text{Slope} = \frac{dy}{dx} = \lim_{h \to 0} \frac{(x+h)^2 - x^2}{h} = \lim_{h \to 0} \frac{2xh + h^2}{h} = \lim_{h \to 0} (2x + h)$$

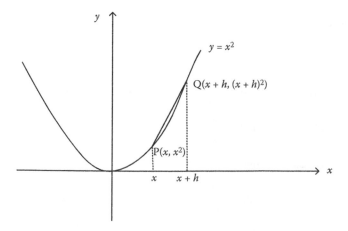

Figure 6.5 Slope of the curve $y = x^2$.

Therefore, as $h \to 0$,

$$\text{Slope} = \frac{dy}{dx} = 2x$$

for the curve $y = x^2$.

This process can be generalized for any function of the form $y = ax^n$. Similar analysis gives

$$\text{Slope} = \frac{dy}{dx} = \lim_{h \to 0} \frac{a(x+h)^n - ax^n}{h}$$

Using the binomial expansion of $(x + h)^n$ as

$$(x+h)^n = x^n + nx^{n-1}h + \frac{n(n-1)}{2!}x^{n-2}h^2 + O(h^3 + higher)$$

$$\text{Slope} = \frac{dy}{dx} = \lim_{h \to 0} \frac{a\left[x^n + nx^{n-1}h + \frac{n(n-1)}{2!}x^{n-2}h^2 + O(h^3 + higher)\right] - ax^n}{h}$$

So,

$$\frac{dy}{dx} = \lim_{h \to 0} a\left[nx^{n-1} + \frac{n(n-1)}{2!}x^{n-2}h + O(h^2 + higher)\right]$$

Now taking the limit as $h \to 0$, the second and higher terms become zero and this gives for any curve given of the form $y = ax^n$ the derivative as

$$\frac{dy}{dx} = anx^{n-1} \tag{6.4}$$

This is now a general formula for finding the slopes of curves for powers of x.

Example 6.2

Given

$$y = x^7, \quad \text{find} \quad \frac{dy}{dx}. \qquad \text{Solution:} \quad \frac{dy}{dx} = 7x^6$$

$$y = 3x^5, \quad \text{find} \quad \frac{dy}{dx}. \qquad \text{Solution:} \quad \frac{dy}{dx} = 15x^4$$

$$y = 2x^{-3}, \quad \text{find} \quad \frac{dy}{dx}. \qquad \text{Solution:} \quad \frac{dy}{dx} = -6x^{-4}$$

$$y = x, \quad \text{find} \quad \frac{dy}{dx}. \qquad \text{Solution:} \quad \frac{dy}{dx} = 1$$

$$y = 1 \quad \text{find} \quad \frac{dy}{dx}. \qquad \text{Solution:} \quad \frac{dy}{dx} = 0$$

Example 6.3

If $y = x^5$, find $\dfrac{dy}{dx}$, then the gradient at the point (2, 32) on this curve.

Solution:

$$\frac{dy}{dx} = 5x^4$$

At the point (2, 32), the gradient is $\dfrac{dy}{dx} = 5 \times 2^4 = 80$.

Note: Finding $\dfrac{dy}{dx}$ *is called* differentiating y with respect to **x**.

What about adding or subtracting powers of x? See the following example.

Example 6.4

If $y = x^6 + 3x^4 - x^2 + x$,   find $\dfrac{dy}{dx}$.

When differentiating, simply add and subtract the separate differentiated terms. So, in this case, $\dfrac{dy}{dx}$ becomes

$$\frac{dy}{dx} = 6x^5 + 12x^3 - 2x + 1$$

6.1.1.1 Differentiating Fractional and Negative Powers of x

The same rule applies as with integer powers of x, that is, using the formula given by Equation 6.4

$$\frac{dy}{dx} = anx^{n-1}$$

applies when the power n is a fraction or a negative number as well.

Example 6.5

If $y = 2x^{\frac{1}{2}} - 6x^{-4}$
then the slope becomes

$$\frac{dy}{dx} = x^{-\frac{1}{2}} + 24x^{-5}.$$

When the function is written in terms of $\sqrt{x}$ or $\dfrac{1}{x^2}$, then to begin with it is probably best to rewrite it with powers that are fractions or negative numbers, then use the rule given by Equation 6.4.

As a reminder, the following are useful when changing to fractional or negative powers:

$$\frac{1}{x^2} = x^{-2} \qquad\qquad \sqrt{x} = x^{\frac{1}{2}}$$

$$\frac{1}{x^3} = x^{-3} \qquad\qquad \frac{1}{\sqrt{x}} = x^{-\frac{1}{2}}$$

$$\frac{1}{x^n} = x^{-n} \qquad\qquad \frac{1}{\sqrt[n]{x^m}} = x^{-\frac{m}{n}}$$

6.1.2 Stationary Points (Maxima and Minima)

Look at the graph shown in Figure 6.6 given by $y = f(x)$. Notice that the highest point on the graph of y against x is reached when $\frac{dy}{dx}$ is zero. This is because the gradient at this point on the graph is zero. So, for such points the x values of the stationary points can be found by *putting the gradient function equal to zero*, and solving this equation. Then the y value can be found from the original equation $y = f(x)$.

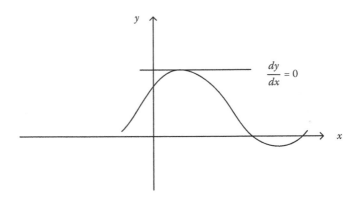

Figure 6.6 Graph of $y = f(x)$ showing stationary points.

Example 6.6

Find the highest point on the graph of $y = 5 + 2x - x^2$.

Solution: First find the gradient function

$$\frac{dy}{dx} = 2 - 2x$$

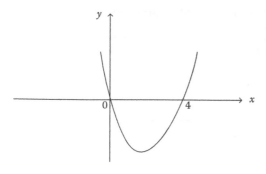

These 'highest' and 'lowest' points are called 'maxima' (single 'maximum') and 'minima' (single 'minimum')

Figure 6.7 Graph showing a minimum point.

At the highest point, the gradient is zero. So,

$$\frac{dy}{dx} = 2 - 2x = 0$$

Solving gives $x = 1$.

To determine the y coordinate, substitute the x value into $y = 5 + 2x - x^2$ to give $y = 6$. So, the highest point on this graph is (1, 6).

The gradient will also be zero if our graph has a lowest point, like the graph of $y = x^2 - 4x$ as sketched in Figure 6.7.

At a maximum or minimum point,

$$\frac{dy}{dx} = 0 \tag{6.5}$$

Is it possible to tell from the equations whether it is a maximum or a minimum point? Clearly, this is important in practical situations; sometimes one wants to maximize profit or minimize wastage and so on.

There are several ways to check out the nature of a *stationary point* (so-called because y is moving neither up nor down at such a point, hence "stationary"). One way is the *practical test* and the other is the *second derivative test*.

6.1.2.1 Practical Test

The practical test looks at what's happening to the gradient on either side of the point. Taking the previous example, $y = 5 + 2x - x^2$ with $\frac{dy}{dx} = 2 - 2x$, you can construct Table 6.1. Here the slope $\frac{dy}{dx}$ is calculated on either side of the stationary point $x = 1$. By looking at the shape of the slope formed, this indicates that the point (1, 6) is a maximum point.

Table 6.1 Considering the Slope on Either Side of the Stationary Point

Value of x	L ($x = 0$)	($x = 1$)	R ($x = 2$)
Sign of $\dfrac{dy}{dx}$	2	0	−2
	╱	────	╲

6.1.2.2   Second Derivative Method

The second derivative method can sometimes be much quicker than the practical test. It involves differentiating twice. Consider what's happening to the gradient as one moves from left to right past a maximum point as shown in Figure 6.8.

So, this can be summarized as

$$\frac{d}{dx}\left(\frac{dy}{dx} \right) = \frac{d^2y}{dx^2} \text{ is negative at a maximum, that is, } \frac{d^2y}{dx^2} < 0 \quad (6.6)$$

On the other hand, consider a minimum point shown in Figure 6.9. This can be summarized as

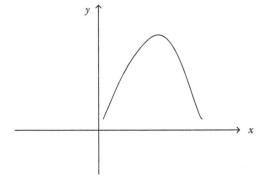

At a local maximum, the gradient is changing from positive to negative, in other words it's decreasing. So the 'gradient of the gradient' is negative.

Figure 6.8 How the gradient function changes at a maximum point.

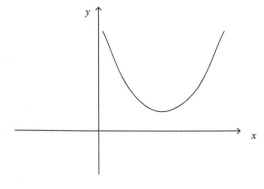

At a local minimum, the gradient is changing from negative to positive, in other words it's increasing. So the 'gradient of the gradient' is positive.

Figure 6.9 How the gradient function changes at a minimum point.

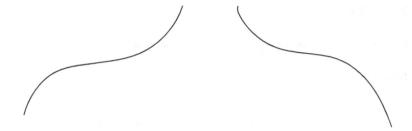

Figure 6.10 Second derivative is zero, showing points of inflection.

$$\frac{d}{dx}\left(\frac{dy}{dx}\right) = \frac{d^2y}{dx^2} \text{ is positive at a minimum, that is, } \frac{d^2y}{dx^2} > 0 \qquad (6.7)$$

What happens if the second derivative is zero? Well, there might be a *point of inflection*, that is, neither a maximum or minimum point, but still the slope is zero as shown in Figure 6.10.

The easiest way to check it out is to make a table, as in the first method of the practical test.

6.1.3 Differentiating Products and Quotients

So far, differentiating powers of x and finding maxima and minima have been considered. But functions such as $y = (1 + 3x)(2x^4 + 6x)$ have not been tackled, at least not without multiplying out the brackets. Next, the derivatives of functions that are products and quotients are considered.

6.1.3.1 Products

This type of function consists of one part multiplied by another; it is a *product* of the two parts. The way we do it is to differentiate only one part at a time.

If $y = (1 + 3x)(2x^4 + 6x)$, then

$$\frac{dy}{dx} = (1 + 3x)(8x^3 + 6) \ + \ (3)(2x^4 + 6x)$$

plus

put down the first part

differentiate the first part

differentiate the second

put down the second

This idea can be expressed mathematically by the *product rule* formula. If $y = u.v$, where u and v are functions of x then

$$\frac{dy}{dx} = u\frac{dv}{dx} + v\frac{du}{dx} \qquad (6.8)$$

Example 6.7

If $y = (x^2 - 2)(x^2 + 1)$, find $\dfrac{dy}{dx}$, simplifying your result.

Solution:

$$\frac{dy}{dx} = (x^2 - 2)(2x) + (2x)(x^2 + 1) = 4x^3 - 2x$$

6.1.3.2 Quotients

Sometimes there may be functions like

$$y = \frac{x^2 + 1}{x^2 - 1}$$

How do you find $\dfrac{dy}{dx}$ in this case?

Well, there's a formula that can be used similar to the one for the product rule. It is a little more complicated than the one for products, but quite simple to use. It is called the *quotient rule*. Expressed mathematically, if $y = \dfrac{u}{v}$, where u and v are functions of x, then

$$\frac{dy}{dx} = \frac{v\dfrac{du}{dx} - u\dfrac{dv}{dx}}{v^2} \tag{6.9}$$

Expressed in words this is,

minus

$$\frac{\text{(put down the bottom part)(differentiate the top)} \quad - \quad \text{(put down the top)(differentiate the bottom)}}{\text{(put down the bottom)}^2}$$

Note: Start with the bottom part; remember it's a minus sign linking the terms on the top. The squared term on the bottom is not usually multiplied out.

Example 6.8

If $y = \dfrac{5x + 1}{5x + 3}$,   find $\dfrac{dy}{dx}$. Simplify your results (top only).

Solution:

$$\frac{dy}{dx} = \frac{(5x+3)(5)-(5x+1)(5)}{(5x+3)^2}$$

$$\frac{dy}{dx} = \frac{10}{(5x+3)^2}$$

6.1.4 Standard Functions

So far, the slope $\frac{dy}{dx}$ when y is some power of x, a product and a quotient have been considered. It is now time to consider how to differentiate such functions as e^{3x}, $\sin x$, and $\cos x$ etc.

The following results can all be proved from first principles, but for the moment it is better to just make use of them. There is no need to learn these, but there will be a need to be able to look them up and use them, both with and without substituting numbers. Table 6.2 shows some standard functions that are commonly used.

Table 6.2 Derivatives of Some Standard Functions

y	$\frac{dy}{dx}$
x^n	nx^{n-1}
e^{kx}	ke^{kx}
$\ln x$	$\frac{1}{x}$
$\sin kx$	$k \cos kx$ (provided x is in radians)
$\cos kx$	$-k \sin kx$ (provided x is in radians)
$\tan kx$	$k \sec^2 kx$ (provided x is in radians)

Examples 6.9

1. If $y = x^2 + 3e^x$, find $\frac{dy}{dx}$ when $x = 2$.

Solution:

$$\frac{dy}{dx} = 2x + 3e^x = 26.2$$

2. If $y = 4x + \sin x$, find $\frac{dy}{dx}$ when $x = 1$.

Solution:

$$\frac{dy}{dx} = 4 + \cos x = 4.54$$

(Be careful! Remember to use radians when working these out.)

Clearly, it's not difficult to differentiate such functions. What about using the rules for product and quotient? They apply just as well to these functions as they did to powers of x. But do make sure the correct rule is being used.

To recap again, there are two arrangements to watch out for they are the product and quotient.

Product: $y = u.v$ then use $\dfrac{dy}{dx} = u\dfrac{dv}{dx} + v\dfrac{du}{dx}$

Quotient: $y = \dfrac{u}{v}$ then use $\dfrac{dy}{dx} = \dfrac{v\dfrac{du}{dx} - u\dfrac{dv}{dx}}{v^2}$

In the following, more examples using the product and quotient rule are given.

Example 6.10

If $y = x^2 e^{3x}$, then

$$\frac{dy}{dx} = (x^2)(3e^{3x}) + (2x)(e^{3x}) = 3x^2 e^{3x} + 2xe^{3x}$$

Example 6.11

If $y = \dfrac{\ln x}{x}$, then

$$\frac{dy}{dx} = \frac{(x)\left(\dfrac{1}{x}\right) - (1)(\ln x)}{x^2} = \frac{1 - \ln x}{x^2}$$

Next is a more complex problem to show how to make use of these derivatives.

Example 6.12

Find any maximum or minimum points on the graph of $y = \dfrac{2x}{e^x}$.

Solution: Remember, the procedure is to first find $\dfrac{dy}{dx}$, then put this equal to zero and solve it to find x. Next find the y-values that correspond with the x-value(s), and last decide which sort of stationary point it is (maximum, minimum, or point of inflection).

(This function is a quotient, so use the quotient rule when differentiating.)

Starting with $y = \dfrac{2x}{e^x}$, find $\dfrac{dy}{dx}$.

$$\frac{dy}{dx} = \frac{(e^x)(2) - (e^x)(2x)}{(e^x)^2} = \frac{2(1-x)}{e^x}$$

At the stationary point, $\dfrac{dy}{dx} = 0$ and so $1 - x = 0$ giving $x = 1$ and $y = \dfrac{2}{e}$.

So, the stationary point is at $\left(1, \dfrac{2}{e}\right)$.

To check what kind of point it is, you could use the second derivative test, but here it is simpler to use the practical test about the point $x = 1$ as shown in Table 6.3.

Table 6.3 Slope of the Curve about the Stationary Point

Value of x	L ($x = 0$)	($x = 1$)	R ($x = 2$)
Sign of $\dfrac{dy}{dx}$	2	0	−0.27

So, the point $\left(1, \dfrac{2}{e}\right)$ is a local maximum point and the graph of this function is shown in Figure 6.11.

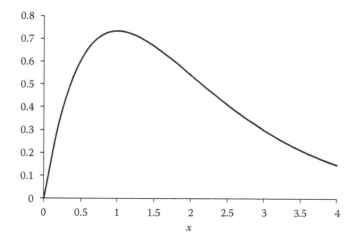

Figure 6.11 Graph of the function $y = \dfrac{2x}{e^x}$.

6.1.5 Function of a Function (Chain Rule)

Consider a function to differentiate that looks like the following: $y = (x^2 + 5x + 1)^9$. This combines two ideas that can be tackled: $y = (x^2 + 5x + 1)$ and $y = x^9$

Think of a function as a box or machine that does something to an input. Then the combined function can be pictured as shown in Figure 6.12.

Clearly, the rate at which the second function changes depends on how fast the first one is changing. To get $\dfrac{dy}{dx}$ for the combined function, just combine the derivatives from each one in a simple way.

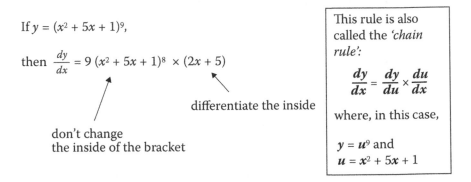

Figure 6.12 Treating the function as a combined function process.

If $y = (x^2 + 5x + 1)^9$,

then $\dfrac{dy}{dx} = 9\,(x^2 + 5x + 1)^8 \times (2x + 5)$

differentiate the inside

don't change
the inside of the bracket

This rule is also called the 'chain rule':

$$\frac{dy}{dx} = \frac{dy}{du} \times \frac{du}{dx}$$

where, in this case,

$y = u^9$ and
$u = x^2 + 5x + 1$

Note: The two parts are multiplied together.

Example 6.13

Find $\dfrac{dy}{dx}$ for the following functions.

1. $y = (4x - 5)^{10}$

Solution: $\dfrac{dy}{dx} = 10(4x - 5)^9(4) = 40(4x - 5)^9$

2. $y = \sin(x^2)$

Solution: $\dfrac{dy}{dx} = \cos(x^2)(2x) = 2x\cos(x^2)$

6.2   Integration

6.2.1   Introduction and the Riemann Sum

What does $\displaystyle\int_a^b f(x)\,dx$ represent? An incomplete answer is to say it is the area under the curve shown in Figure 6.13. Yes, it does give the area under the curve, but the

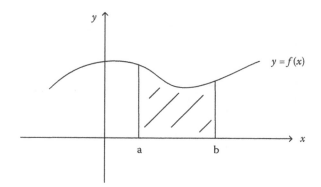

Figure 6.13 Area under the curve $y = f(x)$.

area under the curve is just an application of integration. There are many applications of integration, as will be shown in the later chapters. A more complete answer to the question is as follows: $\int_a^b f(x)\,dx$ is the "limit of approximation by breaking things into pieces." This can be explained further with the diagram in Figure 6.14.

Figure 6.14 An interval on the x axis.

Consider the interval from a to b. Take this interval and chop it up into pieces of varying lengths. Then for each piece pick a point. Then evaluate the function f at these points and then sum over all the pieces multiplied by the width of the piece. So,

$$\int_a^b f(x)\,dx \quad \triangleq \quad \lim_{\Delta x \to 0} \sum_{\text{all pieces}} f(x_p^*)\Delta x \qquad (6.10)$$

This idea can be used in many applications.

Example 6.14

Find the area under the curve $y = f(x)$ from a to b as shown in Figure 6.15.

$$\text{Area} \approx \text{Area of the rectangle} \approx \sum_{\text{all pieces}} f(x_p^*)\Delta x$$

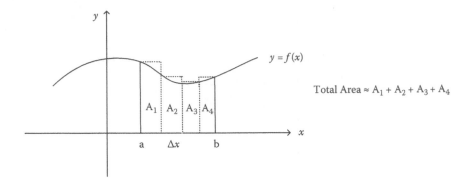

Figure 6.15 Finding the area under the curve in an interval.

By chopping up the interval into smaller and smaller rectangles as $\Delta x \to 0$, the actual area under the curve is obtained.

Therefore,

$$\lim_{\Delta x \to 0} \sum_{all\ pieces} f(x_p^*)\Delta x = \int_a^b f(x)\,dx = \textbf{Area}$$

The curly symbol $\int$ is understood to mean a limiting summation.

6.2.2 Fundamental Theorem of Calculus (Optional Section)

6.2.2.1 How to Compute the Integral $\int_a^b f(x)\,dx$

You will need to make use of the fundamental theorem of calculus, some function f that contains the interval [a, b]. Now define $F(x) = \int_a^x f(t)\,dt$, where x is in the interval [a, b] as shown in Figure 6.16.

Taking the derivative of $F(x)$ as $F(x)'$ gives

$$F(x)' = \lim_{\Delta x \to 0} \left\{ \frac{F(x + \Delta x) - F(x)}{\Delta x} \right\}$$

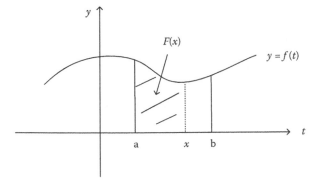

Figure 6.16 Defining an area in a region.

This is from the definition of the limit.
  This can be written as

$$F(x)' = \lim_{\Delta x \to 0} \left\{ \frac{\int_a^{x+\Delta x} f(t)\,dt - \int_a^x f(t)\,dt}{\Delta x} \right\}$$

The area between x and $x + \Delta x$ is shown in Figure 6.17.
  So,

$$\int_a^{x+\Delta x} f(t)\,dt - \int_a^x f(t)\,dt = \text{Shaded area} = \int_x^{x+\Delta x} f(t)\,dt$$

Therefore,

$$F(x)' = \lim_{\Delta x \to 0} \frac{1}{\Delta x} \left\{ \int_x^{x+\Delta x} f(t)\,dt \right\}$$

The *mean value theorem* of definite integrals expresses that there exists a c where $(x \leq c \leq x + \Delta x)$ such that

$$f(c)\Delta x = \text{area under the curve} = \int_x^{x+\Delta x} f(t)\,dt$$

So,

$$f(c) = \frac{1}{\Delta x} \int_x^{x+\Delta x} f(t)\,dt$$

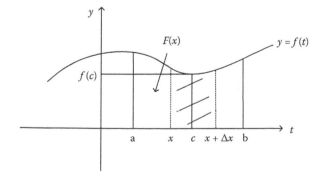

Figure 6.17 Area in the region x and $x + \Delta x$.

Or there exists a c in $[x, x + \Delta x]$ where

$$F(x)' = \lim_{\Delta x \to 0} \{f(c)\}$$

But $x \le c \le x + \Delta x$, so as $\Delta x \to 0$, $c \to x$ and $f(c) \to f(x)$ as $\Delta x \to 0$. Therefore, $F'(x) = f(x)$

Why is this important?

For some function $f(x)$, then there exists a function $F(x)$, thus the derivative of this function is equal to the function $f(x)$. Or, there is an *antiderivative function F(x)* any continuous function $f(x)$. This is also a connection between differentiation and integration.

Now, having this, we can compute $\int_a^b f(x)dx$. This is now given by

$$\int_a^b f(x)\,dx = F(b) - F(a) \qquad (6.11)$$

where $F'(x) = f(x)$.

Since integration is an antiderivative process it then follows that the rule for differentiating any power of x is reversed when for integrating any power of x.

If $y = x^n$, then

$$\int x^n\,dx = \frac{x^{n+1}}{n+1} + C \qquad (6.12)$$

provided that $n \ne -1$ and C is an arbitrary constant.

Note: In words, this particular rule becomes "add one to the power and divide by the new power."

Example 6.15

1. $\int x^4\,dx \;=\; \dfrac{x^5}{5} + C$

2. $\int 3x^5\,dx \;=\; \dfrac{3x^6}{6} + C \;=\; \dfrac{x^6}{2} + C$

3. $\int x^{0.5}\,dx \;=\; \dfrac{x^{1.5}}{1.5} + C$

4. $\int x^{-2}\,dx \;=\; \dfrac{x^{-1}}{-1} + C \;=\; -\dfrac{1}{x} + C$

These are called *indefinite integrals* as the value of the constant C is unknown.

For *definite integrals*, when there are limits of integration, use the formula given by Equation 6.11.

$$\int_a^b f(x)\,dx = F(b) - F(a)$$

Example 6.16

$$\int_0^1 2x\,dx = \left[x^2 + C\right]_0^1 = (1^2 + C) - (0^2 + C) = 1 - 0 = 1$$

Note: The C has canceled out. This will always happen with a definite integral, and so there is no need to write it.

Hence, the preceding calculation would normally be written as

$$\int_0^1 2x\,dx = \left[x^2\right]_0^1 = 1^2 - 0^2 = 1 - 0 = 1$$

6.2.3   Standard Integrals and Areas under Curves

All the integrating so far has been dealing with powers of x.

Just as with differentiating, finding integrals of other functions may be necessary. These are summarized in the Table 6.4.

Table 6.4 Standard Integrals for Some Common Functions

y	$\int y\,dx$
x^n	$\dfrac{x^{n+1}}{n+1} + C, \quad n \neq -1$
$\dfrac{1}{x}$	$\ln x + C$
e^{kx}	$\dfrac{e^{kx}}{k} + C$
$\sin kx$	$-\dfrac{\cos kx}{k} + C$
$\cos kx$	$\dfrac{\sin kx}{k} + C$
$\sec^2 kx$	$\dfrac{\tan kx}{k} + C$
$\tan x$	$-\ln(\cos x) + C$
$\dfrac{1}{x^2 + a^2}$	$\dfrac{1}{a}\tan^{-1}\dfrac{x}{a} + C$

Note: x must be in radians when using the trigonometric functions.

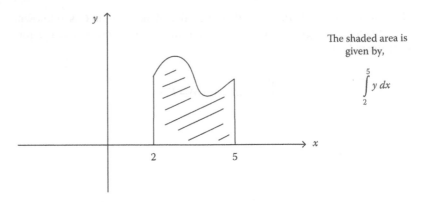

The shaded area is given by,

$$\int_2^5 y\,dx$$

Figure 6.18 Calculating areas under curves.

6.2.3.1   Finding the Area under a Curve

Provided that the whole curve is above the *x*-axis, the definite integral gives the area under a curve as shown in Figure 6.18.

Example 6.17

1. Find the area under the curve $y = 10x - x^2$ between the *x*-values 2 and 8.

Solution: The area will be

$$\int_2^8 (10x - x^2)\,dx = \left[10\frac{x^2}{2} - \frac{x^3}{3}\right]_2^8 = \left\{10\frac{8^2}{2} - \frac{8^3}{3}\right\} - \left\{10\frac{2^2}{2} - \frac{2^3}{3}\right\} = 132$$

So, the required area is 132 square units.

2. Find the area under the curve $y = \sin x$ between the *x*-values 0 and π.

Solution: The area will be

$$\int_0^\pi (\sin x)\,dx = \left[-\cos x\right]_0^\pi = \left\{-\cos \pi\right\} - \left\{-\cos 0\right\} = 1 + 1 = 2$$

So, the required area is 2 square units.

Note: One, or both, of the limits may be negative, as long as the graph stays above the x-axis.

3. Find the area under the curve $y = x^3 + 4x^2 + 3x + 10$ between the *x*-values −3 and −1.

Solution: The area will be

$$\int_{-3}^{-1} (x^3 + 4x^2 + 3x + 10)\, dx = \left[\frac{x^4}{4} + 4\frac{x^3}{3} + 3\frac{x^2}{2} + 10x \right]_{-3}^{-1}$$

$$= \left\{ -9\frac{7}{12} \right\} - \left\{ -32\frac{1}{4} \right\} = 22\frac{2}{3}$$

So, the required area is 22.6667 square units (to four decimal places). This area is shown in Figure 6.19.

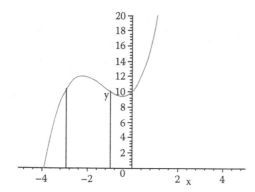

It is found that the area under the graph, between the two vertical lines, is 22.67 square units.

Figure 6.19 Graph of the function $y = x^3 + 4x^2 + 3x + 10$.

What happens if the graph goes below the x-axis? Everything below the axis results in a negative contribution to the integral, so if there are parts above and parts below the axis, then there is a need to work out the upper and lower bits separately. Figure 6.20 shows a cubic function.

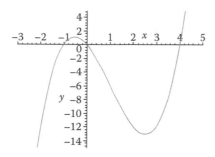

Figure 6.20 Graph of the cubic function $y = x^3 - 3x^2 - 4x$.

4. Find the total area enclosed by the graph and the x-axis between $x = -1$ and $x = 4$.

Solution: First, work out the area above the x-axis between $x = -1$ and $x = 0$:

$$\int_{-1}^{0} (x^3 - 3x^2 - 4x)\, dx = \left[\frac{x^4}{4} - 3\frac{x^3}{3} - 4\frac{x^2}{2} \right]_{-1}^{0} = \{0\} - \left\{ -\frac{3}{4} \right\} = 0.75$$

So, this area is 0.75 square units.

Now work out the area below the x-axis between $x = 0$ and $x = 4$:

$$\int_0^4 (x^3 - 3x^2 - 4x)\,dx = \left[\frac{x^4}{4} - 3\frac{x^3}{3} - 4\frac{x^2}{2}\right]_0^4 = \{-32\} - \{0\} = -32$$

The integral is negative, but the area is positive; so this area is 32 square units.

Adding together these two areas, the total area enclosed by the graph and the x-axis over the interval $-1 \leq x \leq 4$ is 32.75 square units.

6.2.4   Improper Integrals

Recall that for the integral to exist and call it definite, $\int_a^b f(x)\,dx$, then $f(x)$ must be bounded and finite in the interval [a, b].

But sometimes there can be the situation as shown in Figure 6.21. $f(x)$ is a bounded function but not finite, so $\int_a^b f(x)\,dx$ is an improper integral. In Figure 6.22, the interval is finite but $f(x)$ is not bounded. Again, this type of integral shown in Figure 6.22 is called an improper integral.

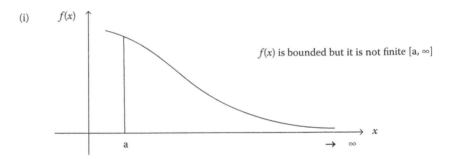

Figure 6.21 $f(x)$ is a bounded function but not finite.

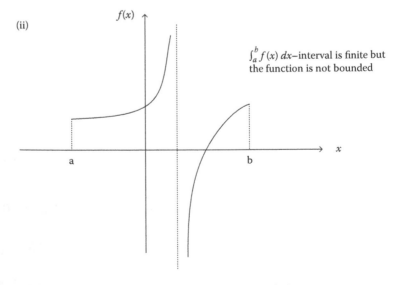

Figure 6.22 The interval is finite but $f(x)$ is not bounded.

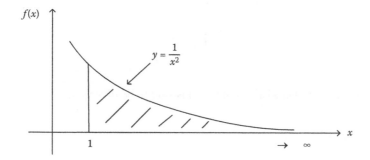

Figure 6.23 The function $y = \dfrac{1}{x^2}$ is finite but unbounded.

Example 6.18

Consider the function given by the curve $y = \dfrac{1}{x^2}$, as shown in Figure 6.23.
Find the area under the curve from $x = 1$ to $x = \infty$.

Solution: This is expressed as

$$A = \int_1^\infty \frac{1}{x^2}\,dx$$

The following describes how to deal with these types of integrals (one of the limits is infinite ∞).
Let x go up to some value $x = t$.

$$\int_1^\infty \frac{1}{x^2}\,dx = \lim_{t \to \infty} \int_1^t \frac{1}{x^2}\,dx$$

Do the integral as normal and at the end use the limit as $t \to \infty$.

$$= \lim_{t \to \infty} \int_1^t \frac{1}{x^2}\,dx = \lim_{t \to \infty} \int_1^t x^{-2}\,dx = \lim_{t \to \infty} \left[\frac{x^{-1}}{-1}\right]_1^t$$

$$= \lim_{t \to \infty} \left[\frac{-1}{x}\right]_1^t = \lim_{t \to \infty} \left[\frac{-1}{t} - \frac{-1}{1}\right] = \lim_{t \to \infty} \left[1 - \frac{1}{t}\right]$$

So, as $t \to \infty$, $\dfrac{1}{t} \to 0$. Therefore, $\displaystyle\int_1^\infty \frac{1}{x^2}\,dx = 1$

Since this number is finite, the integral

$$\int_1^\infty \frac{1}{x^2}\,dx$$

is *convergent*. Otherwise, the integral is said to be *divergent*.

Example 6.19

Calculate $\displaystyle\int_{-\infty}^{+\infty} x^2 e^{-x^3}\,dx$.

This can be split into two equivalent integrals:

$$\int_{-\infty}^{+\infty} x^2 e^{-x^3}\,dx = \int_{-\infty}^{0} x^2 e^{-x^3}\,dx + \int_{0}^{+\infty} x^2 e^{-x^3}\,dx$$

$$\lim_{t \to -\infty} \int_{t}^{0} x^2 e^{-x^3}\,dx + \lim_{v \to \infty} \int_{0}^{v} x^2 e^{-x^3}\,dx$$

Start with

$$\lim_{t \to -\infty} \int_{t}^{0} x^2 e^{-x^3}\,dx = \lim_{t \to -\infty}\left[\frac{-e^{-x^3}}{3}\right]_t^0 = \lim_{t \to -\infty}\left[\frac{-1}{3} + \frac{1}{3}e^{-t^3}\right]$$

As $t \to -\infty$, $e^{-t^3} \to \infty$, so

$$\int_{-\infty}^{0} x^2 e^{-x^3}\,dx \to \infty$$

The first part is divergent and so the original integral is divergent, that is,

$$\int_{-\infty}^{+\infty} x^2 e^{-x^3}\,dx \to \infty$$

Example 6.20

There is an infinite discontinuity in the following integral. There is a problem when $x = 0$ in the integrand.

Calculate $\displaystyle\int_0^9 \frac{1}{\sqrt{x}}\, dx$.

Here, replace the zero with the symbol a and let this $a \to 0$.

$$\lim_{a \to 0^+} \int_a^9 x^{-\frac{1}{2}}\, dx = \lim_{a \to 0^+} \left[2x^{\frac{1}{2}}\right]_a^9 = \lim_{a \to 0^+} \left[2\sqrt{9} - 2\sqrt{a}\right]$$

as $a \to 0$, $\sqrt{a} \to 0$.

Therefore,

$$\int_0^9 \frac{1}{\sqrt{x}}\, dx = 6$$

6.3   Integration Techniques

6.3.1   Substitution

In this method, the idea is to replace the complicated part of the integral by letting it equal a new variable, say, u. This then involves changing the integral over to a u variable problem with respect to du.

Take the first example, $\displaystyle\int \frac{2}{x+3}\, dx$. Write a new letter, say, u, to stand for the expression $x + 3$. Now, one must be very careful to keep track of the three variables x, y and u. The working will look like this after the substitution for u and replacing for dx.

$$\int \frac{2}{x+3}\, dx = \int \frac{2}{u}\, du = 2\ln(u) = 2\ln(x+3) + C$$

$$\begin{aligned} u &= x + 3 \\ \therefore \frac{du}{dx} &= 1 \\ \therefore du &= dx \end{aligned}$$

The reader could check back by differentiating this result and the result is indeed $\dfrac{2}{x+3}$.

Example 6.21

Calculate $\int 4\cos(100\pi x)\,dx$.

First, make the substitution. Let $u = 100\pi x$ and also replacing for dx gives

$$\int 4\cos(100\pi x)\,dx = \int 4\cos(u)\frac{du}{100\pi} = \int \frac{4}{100\pi}\cos(u)\,du$$

$$= \frac{4}{100\pi}\left(-\sin(u)\right) + C$$

$$= -\frac{4}{100\pi}\sin(100\pi x) + C$$

$u = 100\pi x$

$\therefore \dfrac{du}{dx} = 100\pi$

$\therefore \dfrac{du}{100\pi} = dx$

In all of these examples, the connection between du and dx has involved a number. But there might be a situation where one wants to replace $x^2 + 3$, say, with the letter u.

In this case, $\dfrac{du}{dx} = 2x$ and so dx will have to be replaced by $\dfrac{du}{2x}$.

Example 6.22

$$\int \frac{6x}{x^2+3}\,dx$$

Make the substitution. Let $u = x^2 + 3$ and also replacing for dx gives

$$\int \frac{6x}{x^2+3}\,dx = \int \frac{6x}{u}\cdot\frac{du}{2x} = \int \frac{3}{u}\,du$$

$$= 3\ln(u) + C$$

$$= 3\ln(x^2+3) + C$$

$u = x^2 + 3$

$\therefore \dfrac{du}{dx} = 2x$

$\therefore \dfrac{du}{2x} = dx$

Finally, *definite integrals* (those with limits) can be done by the substitution method. Of course, one can do it just as earlier, using the limits at the end to calculate the answer. But it is much quicker to shortcut a little by changing the limits so that one does not return to the variable x at all.

Example 6.23

Consider the integral with limits as

$$\int_1^5 \frac{2x}{x^2+7}\,dx$$

Now, recall that the limits 1 and 5 mean $x = 1$ and $x = 5$.

When changing the variable of integration to u, at the same time the limits of u-values can be changed too. If this is done, there is no need to return to the variable x at all.

But one will have to calculate the two u limits; this can be done in the margin with the '$\dfrac{du}{dx}$' calculation.

So, the whole solution will look like this,

$$\int_{1}^{5} \frac{2x}{x^2 + 7}\,dx = \int_{8}^{32} \frac{2x}{u} \cdot \frac{du}{2x}$$

$$= \int_{8}^{32} \frac{du}{u} = \left[\ln u\right]_{8}^{32}$$

$$= \ln(32) - \ln(8)$$

$$= \ln(4)$$

$$u = x^2 + 7$$

$$\therefore \frac{du}{dx} = 2x$$

$$\therefore \frac{du}{2x} = dx$$

when $x = 1$, $u = 8$

when $x = 5$, $u = 32$

6.3.2 Partial Fractions

If one is faced with an integral that contains a fraction, first check to see if the top is the derivative of the bottom of the fraction. For example:

$$\int \frac{2x}{x^2 + 3}\,dx, \qquad \int \frac{3x^2 + 4x}{x^3 + 2x^2 - 5}\,dx, \qquad \int \frac{x}{x^2 + 4}\,dx$$

Note: In the case of the last example, the exact *of the derivative on the top does not exist, but it can be made to do so as follows,*

$$\int \frac{x}{x^2 + 4}\,dx = \frac{1}{2}\int \frac{2x}{x^2 + 4}\,dx$$

and now this can be done like the others using the method of substitution.

If, however, the top does not contain the derivative of the bottom, and if the bottom part can be factorized, then try using integration using partial fractions.

The different types of forms of partial fractions splits are summarized as follows:

1. If the factors on the bottom are different and linear:

$$\frac{3x + 1}{(x+1)(x+2)(x-3)} = \frac{A}{x+1} + \frac{B}{x+2} + \frac{C}{x-3}$$

2. If there is a repeated factor on the bottom:

$$\frac{1}{(x+2)(x-1)^2} = \frac{A}{x+2} + \frac{B}{x-1} + \frac{C}{(x-1)^2}$$

3. If a factor has higher powers than just x:

$$\frac{3x+1}{(x-1)(x^2+1)} = \frac{A}{x-1} + \frac{Bx+C}{x^2+1}$$

There may be a combination of these types. A, B, and C are constants that need to be worked out before doing the integral.

First, let's try to work out some partial fractions, leaving the integration part for later.

6.3.2.1   Type 1: Different Linear Factors
Example 6.24

Express $\dfrac{5x-1}{(x+1)(x-2)}$ in partial fractions.

1. Write the fraction as separate parts, putting A, B, C, ... for the unknown numerators of the partial fractions. To use the example above, write as follows:

$$\frac{5x-1}{(x+1)(x-2)} = \frac{A}{x+1} + \frac{B}{x-2}$$

2. Make the "little fractions" back up into one "big fraction."

$$\frac{5x-1}{(x+1)(x-2)} = \frac{A}{x+1} + \frac{B}{x-2} = \frac{A(x-2)+B(x+1)}{(x+1)(x-2)}$$

3. Put the tops of the right- and left-hand sides equal. This follows, since the bottoms are equal.

$$5x-1 \equiv A(x-2)+B(x+1)$$

4. Substitute values for x that make each bracket equal to zero. This will give the values for A, B, and so on.

In this case, the first bracket can be made zero if $x = 2$, and the second bracket is zero if $x = -1$.

Letting $x = 2$:

$$5(2)-1 = A(0)+B(2+1)$$
$$\therefore 9 = 3B$$
$$\therefore B = 3$$

Letting $x = -1$:

$$5(-1) - 1 = A(-1 - 2) + B(0)$$
$$\therefore -6 = -3A$$
$$\therefore A = 2$$

Now, the partial fraction split has finished. Next, write the partial fraction out as

$$\frac{5x - 1}{(x + 1)(x - 2)} = \frac{2}{x + 1} + \frac{3}{x - 2}$$

6.3.2.2 Type 2: Denominator with a Repeated Factor
Example 6.25

Express $\dfrac{1}{(x + 2)(x - 1)^2}$ in partial fractions.

First, write the form of the partial fraction, referring to different forms:

$$\frac{1}{(x + 2)(x - 1)^2} = \frac{A}{x + 2} + \frac{B}{x - 1} + \frac{C}{(x - 1)^2}$$

Now proceed as before. First, make the right-hand side into one big fraction:

$$\frac{1}{(x + 2)(x - 1)^2} = \frac{A}{x + 2} + \frac{B}{x - 1} + \frac{C}{(x - 1)^2} = \frac{A(x - 1)^2 + B(x + 2)(x - 1) + C(x + 2)}{(x + 2)(x - 1)^2}$$

Then make the top parts equal, since the bottom parts are the same:

$$1 = A(x - 1)^2 + B(x + 2)(x - 1) + C(x + 2)$$

And now need to think of which numbers we could substitute for x that will make the brackets equal to zero. If $x = 1$, then

$$1 = A(0)^2 + B(2)(0) + C(3)$$

So, $1 = 3C$, which gives $C = \dfrac{1}{3}$.
If $x = -2$, then

$$1 = A(-3)^2 + B(0)(-3) + C(0)$$

So, $1 = 9A$, which gives $A = \dfrac{1}{9}$.

Now, to find B, either substitute any other number for x and make use of the values of A and C, or look at the powers of on each side of the equation to equate coefficients of the powers of x.

Start with

$$1 = A(x-1)^2 + B(x+2)(x-1) + C(x+2)$$

Looking at the coefficients of

$$[x^2]: \qquad 0 = A + B = \frac{1}{9} + B$$

$$\therefore B = -\frac{1}{9}$$

The work is done; just write out the partial fraction in full:

$$\frac{1}{(x+2)(x-1)^2} = \frac{\frac{1}{9}}{x+2} + \frac{-\frac{1}{9}}{x-1} + \frac{\frac{1}{3}}{(x-1)^2}$$

This is more useful written with the constants at the front as follows:

$$\frac{1}{(x+2)(x-1)^2} = \frac{1}{9}\left(\frac{1}{x+2}\right) - \frac{1}{9}\left(\frac{1}{x-1}\right) + \frac{1}{3}\left(\frac{1}{(x-1)^2}\right)$$

6.3.2.3 Type 3: Denominator with a Quadratic Factor
Example 6.26

Express $\dfrac{3x+1}{(x-1)(x^2+1)}$ in partial fractions.

Start by writing the correct form for the partial fraction, referring to the list of types:

$$\frac{3x+1}{(x-1)(x^2+1)} = \frac{A}{x-1} + \frac{Bx+C}{x^2+1}$$

Now proceed as before, first making the right-hand side into one big fraction:

$$\frac{3x+1}{(x-1)(x^2+1)} = \frac{A}{x-1} + \frac{Bx+C}{x^2+1} = \frac{A(x^2+1)+(Bx+C)(x-1)}{(x-1)(x^2+1)}$$

Now putting the top parts equal, since the bottom parts are the same:

$$3x+1 = A(x^2+1) + (Bx+C)(x-1)$$

This time, there will only be one number that can be substituted for x to obtain A.

If $x = 1$, then

$$3(1)+1 = A(2)+(Bx+C)(0)$$
$$\therefore 4 = 2A$$
$$\therefore A = 2$$

Check the constant terms by putting $x = 0$, $1 = A + C(-1)$, so $C = 1$. Finally, to get a value for B, look at the coefficients of x^2:

$$[x^2]: \qquad 0 = A + B$$
$$\therefore B = -2$$

All three numbers A, B, and C are found and now the partial fraction can be written out in full as

$$\frac{3x+1}{(x-1)(x^2+1)} = \frac{2}{x-1} + \frac{-2x+1}{x^2+1}$$

6.3.2.4 Performing the Final Integration

Usually, will end up with three different types of function to integrate as follows:

1. $\int \dfrac{3}{x+1}\,dx$

2. $\int \dfrac{2}{(x+4)^2}\,dx$

3. $\int \dfrac{2x+5}{x^2+1}\,dx$

The first two can be done by substitution, although the results are different. The last integral has to be tackled in a different way. Another standard integral must be used, which again can be looked up in a standard table of integrals as shown in Table 6.2 and is reproduced here.

$$\int \frac{1}{x^2+a^2}\,dx = \frac{1}{a}\tan^{-1}\frac{x}{a} + C \tag{6.13}$$

Again looking at the last integral from earlier,

$$\int \frac{2x+5}{x^2+1}\,dx$$

what must be done here is split the top of the fraction and make two parts out of the expression as follows:

$$\int \frac{2x+5}{x^2+1} dx = \int \frac{2x}{x^2+1} dx + \int \frac{5}{x^2+1} dx$$

The first part can be done by substitution again; the second part can be done using the standard integral given by Equation 6.13, where $a^2 = 1$, and so $a = 1$. Putting all this together gives the solution as

$$\int \frac{2x+5}{x^2+1} dx = \int \frac{2x}{x^2+1} dx + \int \frac{5}{x^2+1} dx = \ln(x^2+1)) + 5\tan^{-1} x + C$$

6.3.3   Integration by Parts

Sometimes there is a product of two functions that needs integrating as follows:

$$\int x\, e^x\, dx$$

There is a formula that is derived from the product rule and can be used to find the integral of a product of two functions, that is,

$$\int u.\frac{dv}{dx} dx = u.v - \int v.\frac{du}{dx} dx \qquad (6.14)$$

To use this formula, the term, which is $\frac{dv}{dx}$, must be able to find its integral easily. Now, if both terms are easy to integrate, then any powers of x is chosen to be the u term.

Example 6.27

Evaluate $\int x\, e^x\, dx$

Since both functions are easy to integrate let $u = x$ and then $\frac{dv}{dx} = e^x$.

For the right-hand side of Equation 6.14, you need the $\frac{du}{dx}$ and the v terms.

$$\frac{du}{dx} = 1 \quad \text{and} \quad v = e^x$$

so,

$$\int xe^x\, dx = xe^x - \int e^x\, dx$$

This gives the final answer as

$$\int xe^x \, dx = xe^x - e^x + C = (x-1)e^x + C$$

Example 6.28

Evaluate $\int x \ln(x) \, dx$

Now clearly the $\ln(x)$ is not easy to integrate so it must be the u term, that is, $u = \ln(x)$ and then $\dfrac{dv}{dx} = x$.

$$\frac{du}{dx} = \frac{1}{x} \text{ and } v = \frac{x^2}{2}$$

Using the formula given by Equation 6.14 gives the solution as

$$\int x \ln(x) \, dx = \frac{x^2}{2} \ln(x) + \frac{x^2}{4} + C$$

Remember, the use of integration to find areas 'under' curves, and, as an extension of this, areas 'between two graphs'.

In general, it's easy to find the area enclosed between two graphs by using the formula,

$$\text{Area} = \int_a^b \left\{ \left(Upper\ graph \right) - \left(Lower\ graph \right) \right\} dx \qquad (6.15)$$

Example 6.29

Review Figure 6.24. To find the bounded area between the curves as shown in Figure 6.24, simply find the points of intersection of the two curves by equating the y values as follows:

$$4 - x^2 = x^2 - 2x$$

$$x^2 - x - 2 = 0$$

so $x = -1$ and $x = 2$.

Performing the integration as given by Equation 6.15 gives

$$\text{Area} = \int_{-1}^2 (4 - x^2) - (x^2 - 2x) \, dx = \int_{-1}^2 (-2x^2 + 2x + 4) \, dx = 9 \text{ square units.}$$

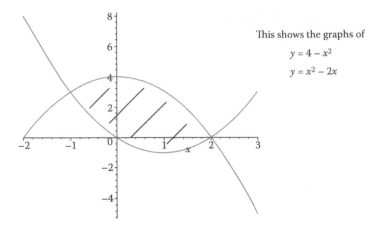

This shows the graphs of

$$y = 4 - x^2$$
$$y = x^2 - 2x$$

Figure 6.24 Area bounded between two curves.

6.4   Applications

Example 6.30: Practical Optimization Problem

In some real practical situations, there may be some maximization or minimization problems. Consider the following example of constructing an open cylindrical tank using minimum material.

An open tank that has vertical sides and a circular base is to be constructed from metal so as to use the minimum amount of material. If the capacity of the tank is to be 8 m³, find the dimensions of the tank shown in Figure 6.25.

Solution: The volume of the tank gives πr2h = 8. The surface area of the tank is given as S = πr2 + 2πrh. You need to find the value of r (and h), which makes S a minimum.

From the first equation, h can be defined as

$$h = \frac{8}{\pi r^2}$$

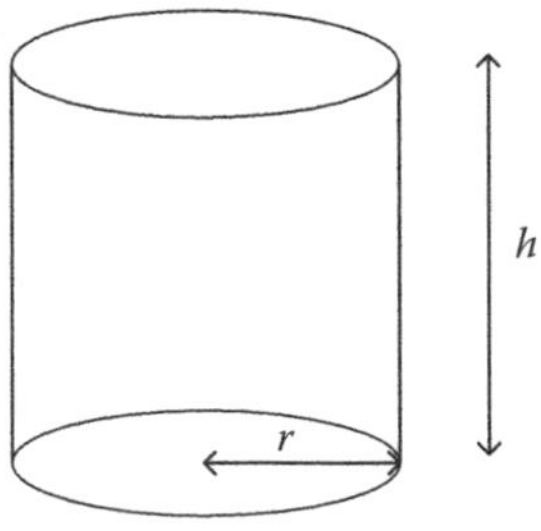

Base radius r, height h

h

r

Figure 6.25 Open tank with radius r and height h.

Substituting this into the equation for S gives

$$S = \pi r^2 + \frac{16}{r}.$$

Now the problem is one of minimizing S with respect to r. Find $\dfrac{dS}{dr} = 0$. Differentiating S with respect to r gives

$$\frac{dS}{dr} = 2\pi r - \frac{16}{r^2} = 0$$

Solving for r gives

$$r^3 = \frac{8}{\pi}$$

which gives

$$r = \left(\frac{8}{\pi}\right)^{\frac{1}{3}} = 1.37$$

This then gives h as

$$h = \frac{8}{\pi(1.37)^2}$$

which gives $h = 1.37$ m.

To check this gives a minimum value of S, not a maximum. Using the second derivative method gives

$$\frac{d^2S}{dx^2} = 2\pi - 16(-2)r^{-3} = 2\pi + \frac{32}{r^3}$$

When $r = 1.37$ m, this is a positive number $= 18.7$. So S is a minimum as required.

Example 6.31: Heat Released during a Fire

An application of how integration can be used in fire engineering is the development of an initial fire growth curve where $\dot{Q}(t)$ is the heat release rate and t is the time in seconds, as shown in Figure 6.26. This represents the initial fire growth.

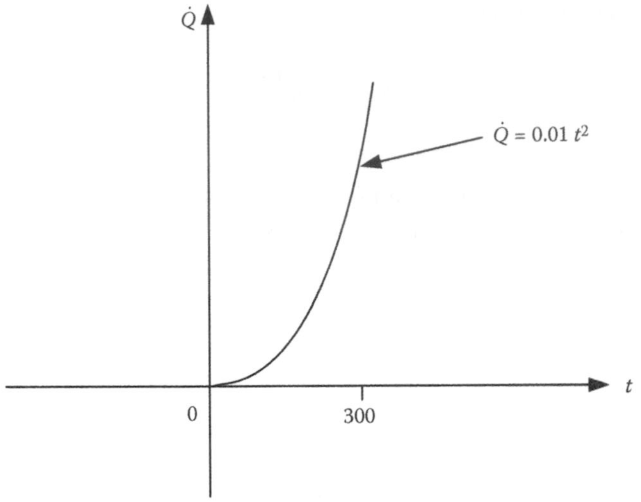

Figure 6.26 Heat release rate $\dot{Q}(t)$ against time t.

Solution: The heat-released Q in the time interval from $t = 0$ to $t = a$ seconds is given by the formula

$$Q = \int_0^a \frac{dQ}{dt}\, dt \qquad (6.16)$$

To calculate the heart released in the first 300 seconds of a fire with the heat-release rate given as a t-squared function, $\dot{Q}(t) = 0.01t^2$, using the formula given in Equation 6.16 gives

$$Q = \int_0^{300} 0.01\, t^2\, dt = \left[\frac{0.01\, t^3}{3} \right]_0^{300} = \frac{(0.01)(300)^3}{3} = \frac{270000}{3} = 90000\, J = 90\, KJ$$

Example 6.32: Applications in Reliability Theory

System or component failure can have a devastating impact on a human level and also on a financial level, so it is important to be able to predict when things are going to fail. The reliability of a component (or system) at time t, say, $R(t)$, is defined in terms of probability as $R(t) = P(T > t)$, where T is the lifetime of the component. This is interpreted as the component is still functioning at time t. $R(t)$ is called the reliability function.

If $f(t)$ is the probability density function (pdf) of the failure function T, then the *probability of failure* up to a time $T \leq t$ can be written as

$$P(T \leq t) = \int_0^t f(s)\, ds,$$

Now, the reliability of a component means the probability that the system has not failed in the interval $[0, t]$. So,

$$R(t) = 1 - P(T \leq t)$$

and in terms of calculations, the reliability $R(t)$ becomes

$$R(t) = P(T > t) = 1 - \int_0^t f(s)\,ds = \int_t^\infty f(s)\,ds$$

It can be seen that integrals can play an important role in calculating physical quantities that can be used to predict system behavior.

Another important application in the reliability of systems is that given the failure probability density function $f(t)$, then the expected time to failure $E(T)$ can be calculated using the formula

$$E(T) = \int_0^\infty t\, f(t)\,dt$$

Now, since $f(t)$ is some function of time and so is t, this integral will be a product of two functions of time. For certain functions of $f(t)$, this will need to be done using the integration by parts method discussed in the earlier sections.

The expected value is a measure of when something will typically fail. This is an important measure as this allows for maintenance of systems to take place in advance of this time, which will prevent a system from failing and can also save lives and money.

Problems

6.1 Find $\dfrac{dy}{dx}$ for each of the following:

a. $y = 4x^3 - 5x^2 + 7x - 11$

b. $y = 8x(\sqrt{x} + 5)$

c. $y = x^2 e^{3x}$

d. $y = \dfrac{\sin x}{x}$

e. $y = (x^3 + 11)^5$

6.2 Find the coordinates of the stationary points for the curve with equation

$$y = 8 - x^2 - \frac{16}{x^2}$$

Determine the nature of the stationary points.

6.3 A glass window consists of a rectangle with sides of length $2r$ cm by h cm and a semicircle of radius r cm as shown in Figure 6.27. The total area of one surface of the glass is 500 cm².

a.   Show that the perimeter P of the window is given by

$$P = \left(2 + \frac{\pi}{2}\right)r + \frac{500}{r}$$

b.   Determine the value of r for which P has a stationary value and hence determine its nature.

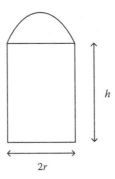

h

$2r$

Figure 6.27 Glass window made from a rectangle and a semicircle.

6.4 Evaluate each of the following:

a. $\displaystyle\int (4x^3 + 9x^2 - 10x + 4)\,dx$

b. $\displaystyle\int \sqrt{x}\,(3x + 5)\,dx$

c. $\displaystyle\int \frac{x^5 + 1}{x^2}\,dx$

d. $\displaystyle\int 6x\sqrt{x^2 + 5}\,dx$

e. $\displaystyle\int_5^6 \frac{2x+1}{(x-4)(x+2)}\,dx$

f. $\displaystyle\int x\,e^{4x}\,dx$

g. $\displaystyle\int e^x \sin x\,dx$

h. $\displaystyle\int x^n \ln x\,dx$

6.5 Given that the initial fire growth is governed by the heat release rate $\dot{Q}(t) = 0.3t^2$, determine the heat released by the fire during the first 10 seconds of burning.

6.6 Given the failure probability density function $f(t)$, the time t being measured in operator work years as follows:

$$f(t) = \begin{cases} \dfrac{1}{32}(t-1), & 1 \leq t \leq 9 \\ 0, & \text{otherwise} \end{cases}$$

Calculate the expected time to failure $E(t)$.

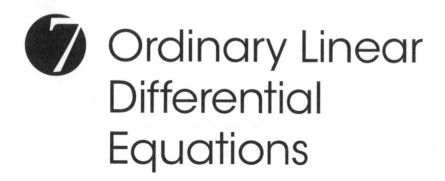 Ordinary Linear Differential Equations

7.1 Background

In many physical situations, equations arise that involve differential coefficients such as $\dfrac{dy}{dx}$, $\dfrac{d^2y}{dx^2}$, $\dfrac{dy}{dt}$ etc. These equations are called differential equations since they contain differential coefficients. Differential equations arise naturally in the modeling of real phenomena in engineering. The following example shows some different areas in which they can occur.

Example 7.1

1. If a body, for example, a sprinkler droplet, falls freely under gravity g, the distance traveled s is given by using Newton's second law of motion as

$$\frac{d^2s}{dt^2} = g \tag{7.1}$$

where s is the distance fallen in time t.

2. Consider the electrical RL circuit shown in Figure 7.1. An electrical circuit with a resistor R and an inductor L in series, the current i flowing in the circuit is given by using Kirchhoff's voltage law as

$$L\frac{di}{dt} + Ri = E \tag{7.2}$$

where i is the current at time t.

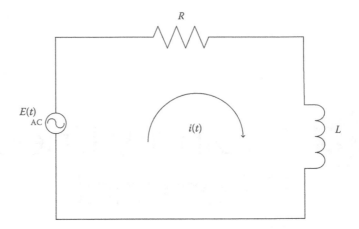

Figure 7.1 A simple *RL* electrical circuit.

3. When a mass oscillates on the end of a spring and is subject to a frictional resistance proportional to its speed, the equation of motion may be written as

$$m\frac{d^2x}{dt^2} + r\frac{dx}{dt} + sx = 0 \qquad (7.3)$$

where x is the displacement from the equilibrium position at time t; and m, r, and s are constants.

Note: In all of these cases the problem is to find the dependent variable in terms of the independent one. For example, from Equations 7.1 to 7.3, respectively, find s in terms of t, find i in terms of t, or find x in terms of t.

The formulation of a mathematical equation to represent a physical situation is referred to as *mathematical modeling*. Differential equations being a big subject area, it is important to define some of the important terminology associated with them in the next section.

7.2   Types of Differential Equations

7.2.1   Introduction

An equation in which at least one term is a differential coefficient is called a *differential equation*. Ordinary differential equations (ODEs) involve only one independent variable x and a dependent variable y, and one or more differential coefficients. Some further examples of these are given next.

Example 7.2

$$x\frac{dy}{dx} = 3y \qquad (7.4)$$

Example 7.3

$$\frac{d^2y}{dx^2} + 2\frac{dy}{dx} + y = \sin x \tag{7.5}$$

Differential equations represent dynamical relationships, that is, quantities that change and so are found to occur in many scientific and engineering problems, and are essential to the study of transient and nonsteady system behavior.

7.2.2   Order of a Differential Equation

The order of a differential equation is given by the highest derivative found in the equation.

What is the order of the differential equations given in Example 7.2 and Example 7.3? Example 7.2 is an example of a first-order differential equation because the highest derivative is $\frac{dy}{dx}$. Example 7.3 is an example of a second-order differential equation because the highest derivative is $\frac{d^2y}{dx^2}$.

What is the order of the differential equations given in the following examples?

Example 7.4

$$\frac{d^3y}{dx^3} + 8y^2\frac{d^2y}{dx^2} + 2\frac{dy}{dx} + y = 3 \tag{7.6}$$

The order is 3 because the highest derivative is the third derivative $\frac{d^3y}{dx^3}$ term.

Example 7.5

$$\left(\frac{dy}{dx}\right)^3 + y^2 = x \tag{7.7}$$

The order is 1 because the highest derivative is the first derivative $\frac{dy}{dx}$ term.

7.2.3   Degree of a Differential Equation

The degree of a differential equation is the power to which the highest derivative is raised.

In Example 7.5, Equation 7.7 has order 1 but the degree is 3, since the highest derivative $\frac{dy}{dx}$ is being raised to the power 3.

7.2.4   Linearity

A differential equation is linear if it is linear in the dependent variable y and its derivatives. The differential equations in Examples 7.2 and 7.3,

$$x\frac{dy}{dx} = 3y \quad \text{and} \quad \frac{d^2y}{dx^2} + 2\frac{dy}{dx} + y = \sin x$$

are both linear.

But the differential equations given by Example 7.4 and 7.5,

$$\frac{d^3y}{dx^3} + 8y^2\frac{d^2y}{dx^2} + 2\frac{dy}{dx} + y = 3 \quad \text{and} \quad \left(\frac{dy}{dx}\right)^3 + y^2 = x$$

are both nonlinear because of the $y^2\dfrac{d^2y}{dx^2}$ and $\left(\dfrac{dy}{dx}\right)^3$ terms being present.

Consider the differential equation given in the following example.

Example 7.6

$$x^3\frac{dy}{dx} + 4\frac{dy}{dx} + y = x^5 \tag{7.8}$$

This equation is still linear since it is linear in y and its derivatives the x powers are not a problem, because x is the independent variable not the dependent variable.

7.2.5 What Is Meant by Solving Differential Equations?

A differential equation represents a relationship between two variables. The same relationship can often be expressed in a form that does not contain the differential coefficient. The following example illustrates this idea.

Example 7.7

If $\dfrac{dy}{dx} = 2x \quad \Rightarrow \quad y = x^2 + C$

These expressions are the same relationship in different forms. Converting a differential equation into a direct equation between y and x is called solving the differential equation.

Note: When solving a first-order differential equation, the solution will contain only one arbitrary constant. A second-order differential equation will produce a solution with two arbitrary constants, and so on.

7.3 First-Order Differential Equations

When dealing with first-order differential equations, the task is to see what type of method is most appropriate to deal with the problem. The next sections consider three methods that can be used to solve certain first-order differential equations, starting with the simplest case, which is using direct integration.

7.3.1 Simplest Situation

If the differential equation can be arranged in the form $\dfrac{dy}{dx} = f(x)$, then this type of equation can be solved by *direct integration* and there is generally no complications involved. The following example shows this simplest case.

Example 7.8

Solve

$$x\frac{dy}{dx} = 6x^3 + 7 \tag{7.9}$$

Solution: Dividing Equation 7.9 by x gives

$$\frac{dy}{dx} = 6x^2 + \frac{7}{x} \tag{7.10}$$

Integrating both sides gives

$$\int \frac{dy}{dx}\,dx = \int \left(6x^2 + \frac{7}{x}\right) dx \tag{7.11}$$

$$y = 2x^3 + 7\ln x + C \tag{7.12}$$

Note: Equation 7.12 *is called a general solution to the differential equation because of the arbitrary constant C present in the solution.*

To obtain the constant C, extra information is needed relating x and y. Once the value of C is known, then the *particular solution* is said to be obtained.

7.3.2 Separating Variables

If the differential equation is not of the simple type, that is, $\dfrac{dy}{dx} = f(x)$, then it may be the case that the differential equation is of the form $\dfrac{dy}{dx} = f(x,y)$. Then the variable y on the right-hand side prevents solving by direct integration.

In this case, the function $f(x,y)$ can be split into two separate functions as follows:

- $F(x)$, a function containing just x terms

- $G(y)$, a function containing just y terms

It may be the case that $f(x,y)$ can be made into a separable form as follows:

$$\frac{dy}{dx} = f(x,y) = F(x).G(y) \tag{7.13}$$

or

$$\frac{dy}{dx} = f(x,y) = \frac{F(x)}{G(y)} \tag{7.14}$$

If the differential equation can be written as either Equation 7.13 or 7.14, then it is possible to separate the variables x and y, and then rearrange the equations to integrate separately as shown in the following examples.

Example 7.9

Solve

$$\frac{dy}{dx} = \frac{2x}{y+1} \tag{7.15}$$

This equation is of the form

$$\frac{dy}{dx} = f(x,y) = \frac{F(x)}{G(y)} \tag{7.16}$$

Solution: Multiplying Equation 7.15 by the $(y + 1)$ term gives

$$(y+1)\frac{dy}{dx} = 2x \tag{7.17}$$

Equation 7.17 is now of the form in which the variables are separated. Now integrating both sides of Equation 7.17 with respect to x gives

$$\int (y+1)\frac{dy}{dx}\,dx = \int 2x\,dx \tag{7.18}$$

$$\int (y+1)\frac{dy}{\cancel{dx}}\,\cancel{dx} = \int 2x\,dx \tag{7.19}$$

$$\int (y+1)\,dy = \int 2x\,dx \tag{7.20}$$

Equation 7.20 can now be integrated separately on each side to give

$$\frac{y^2}{2} + y = x^2 + C \tag{7.21}$$

This then the general solution to the differential equation given by Equation 7.15.

Example 7.10

Suppose we require the equation of a curve that satisfies the following differential equation:

$$2\frac{dy}{dx} = \frac{\cos x}{y} \tag{7.22}$$

and passes through the point (0, 2).

Solution: Separating the variables by multiply Equation 7.22 by y and integrating both sides gives

$$\int 2y\frac{dy}{dx}dx = \int \cos x\, dx \tag{7.23}$$

$$\int 2y\frac{dy}{dx}dx = \int \cos x\, dx \tag{7.24}$$

$$\int 2y\, dy = \int \cos x\, dx \tag{7.25}$$

$$y^2 = \sin x + C \tag{7.26}$$

Now at the point (0, 2) substituting in $x = 0$ and $y = 2$ into Equation 7.26 gives

$$4\sin 0 + C \Rightarrow C = 4$$

So the particular solution becomes

$$y^2 = \sin x + 4 \tag{7.27}$$

Example 7.11

Solve

$$\frac{dy}{dx} = (1+x)(1+y) \tag{7.28}$$

Solution: Dividing Equation 7.28 by $(1 + y)$ separates the variables then integrating both sides gives

$$\int \frac{1}{(1+y)}\frac{dy}{dx}dx = \int (1+x)dx \tag{7.29}$$

$$\int \frac{1}{(1+y)} \frac{dy}{dx} dx = \int (1+x) dx \qquad (7.30)$$

$$\int \frac{1}{(1+y)} dy = \int (1+x) dx \qquad (7.31)$$

Integrating both sides of Equation 7.31 gives the solution as

$$\ln(1+y) = x + \frac{x^2}{2} + C \qquad (7.32)$$

This is the general solution to the differential Equation 7.28.

Example 7.12

Solve

$$\frac{dy}{dx} = \frac{y^2 + xy^2}{x^2 y - x^2} \qquad (7.33)$$

Separating the variables requires more work in this problem. First a common factor of y^2 on the numerator and x^2 on the denominator of Equation 7.33 can be factored out to give

$$\frac{dy}{dx} = \frac{y^2(1+x)}{x^2(y-1)} \qquad (7.34)$$

To separate the variables, multiply Equation 7.34 by $\frac{(y-1)}{y^2}$ on both sides and then integrating gives

$$\int \frac{(y-1)}{y^2} \frac{dy}{dx} dx = \int \frac{(1+x)}{x^2} dx \qquad (7.35)$$

$$\int \frac{(y-1)}{y^2} \frac{dy}{dx} dx = \int \frac{1+x}{x^2} dx \qquad (7.36)$$

Dividing out on both sides of Equation 7.36 by the denominators gives

$$\int \left(\frac{y}{y^2} - y^{-2} \right) dy = \int \left(x^{-2} + \frac{x}{x^2} \right) dx \qquad (7.37)$$

$$\int \left(\frac{1}{y} - y^{-2} \right) dy = \int \left(x^{-2} + \frac{1}{x} \right) dx \qquad (7.38)$$

Integrating both sides of Equation 7.38 gives

$$\ln y + y^{-1} = -y^{-1} + \ln x + C \qquad (7.39)$$

Tidying up both sides gives the solution of the differential Equation 7.33 as

$$\ln y + \frac{1}{y} = \ln x - \frac{1}{x} + C \qquad (7.40)$$

Note: When trying to solve first-order differential equations, it may be the case that the differential equation cannot be solved by separating the variables. In this case another approach is needed to solve the differential equation. The next section considers the method of using the integrating factor technique.

7.3.3 Integrating Factor Technique

If the differential equations have either of the following structures:

$$\frac{dy}{dx} = f(x) \qquad (7.41)$$

This type can be solved using direct integration. Or of the form,

$$\frac{dy}{dx} = F(x).G(y) \qquad (7.42)$$

This type can be solved using separating variables method.

However, if the differential equation is not of the form given by Equation 7.41 or 7.42, then it may be possible to see if the differential equation is of the following form:

$$\frac{dy}{dx} + P(x)y = Q(x) \qquad (7.43)$$

where P and Q are functions of x.

Equation 7.43 cannot generally be solved using the method of separating variables. Here the aim is to try to make the left-hand side of Equation 7.43 into a *complete differential coefficient* from which we can integrate easily. That is to see if Equation 7.43 can be written as

$$\frac{d}{dx}(\ldots\ldots) = Q$$

To do this, Equation 7.43 has to be multiplied by a factor called an *integrating factor* (I.F.), which turns out to be

$$\text{I.F.} = e^{\int P(x)\,dx} \tag{7.44}$$

If Equation 7.42 is multiplied by the I.F., then it becomes

$$\frac{dy}{dx}e^{\int P(x)\,dx} + Pye^{\int P(x)\,dx} = Qe^{\int P(x)\,dx} \tag{7.45}$$

Now the left-hand side of Equation 7.45 is the differential coefficient of $ye^{\int P(x)\,dx}$, so Equation 7.45 can now be written as

$$\frac{d}{dx}\left(ye^{\int P(x)\,dx}\right) = Qe^{\int P(x)\,dx} \tag{7.46}$$

Integrating both sides of Equation 7.46 gives

$$ye^{\int P(x)\,dx} = \int Qe^{\int P(x)\,dx}\,dx \tag{7.47}$$

The solution to the differential of the form given by Equation 7.43 can be written in a form that is easier to remember as follows:

$$y(\text{I.F.}) = \int Q(\text{I.F.})\,dx \tag{7.48}$$

where

$$\text{I.F.} = e^{\int P(x)\,dx} \tag{7.49}$$

Example 7.13

Solve

$$\frac{dy}{dx} + 5y = e^{2x} \tag{7.50}$$

Solution: Comparing with the standard form equation,

$$\frac{dy}{dx} + P(x)y = Q(x) \tag{7.51}$$

implies $P(x) = 5$ and $Q(x) = e^{2x}$.

$$\text{I.F.} = e^{\int P(x)\,dx} = e^{\int 5\,dx} = e^{5x}$$

This gives the I.F. as

$$\text{I.F.} = e^{5x} \tag{7.52}$$

Using Equation 7.47 gives the solution as

$$ye^{5x} = \int e^{2x} e^{5x}\,dx \tag{7.53}$$

Simplifying the exponents gives

$$ye^{5x} = \int e^{7x}\,dx \tag{7.54}$$

$$ye^{5x} = \frac{e^{7x}}{7} + C \tag{7.55}$$

Now dividing both sides of Equation 7.55 by e^{5x} gives the final solution as

$$y = \frac{e^{2x}}{7} + Ce^{-5x} \tag{7.56}$$

Example 7.14

Solve

$$(x+1)\frac{dy}{dx} + y = (x+1)^2 \tag{7.57}$$

Solution: Compare with the standard form

$$\frac{dy}{dx} + P(x)y = Q(x) \tag{7.58}$$

Equation 7.57 is not in standard form and needs dividing by $(x + 1)$ throughout to give

$$\frac{dy}{dx} + \frac{1}{x+1}y = (x+1) \tag{7.59}$$

This implies that $P(x) = \dfrac{1}{x+1}$ and $Q(x) = x + 1$.

$$\text{I.F.} = e^{\int P(x)dx} = e^{\int \frac{1}{x+1}dx} = e^{\ln(x+1)} = x + 1$$
$$\therefore \quad \text{I.F.} = x + 1 \tag{7.60}$$

Using Equation 7.47 gives the solution as

$$y(x+1) = \int (x+1)^2\, dx \tag{7.61}$$

Integrating the right-hand side of Equation 7.61 gives

$$y(x+1) = \frac{(x+1)^3}{3} + C \tag{7.62}$$

Dividing both sides of Equation 7.62 by $(x + 1)$ gives the solution as

$$y = \frac{(x+1)^2}{3} + \frac{C}{x+1} \tag{7.63}$$

7.4   Second-Order Differential Equations

In the previous section, the differential equations considered had in them the highest derivative of order 1. Now more complicated differential equations are considered where the highest derivative is of order 2. These are called second-order differential equations. The general second-order linear differential equation is of the following form:

$$a(x)\frac{d^2y}{dx^2} + b(x)\frac{dy}{dx} + c(x)y = f(x) \tag{7.64}$$

where $a(x)$, $b(x)$, $c(x)$, and $f(x)$ are generally functions of x.

Alternatively, a shorthand notation is used to represent the derivative functions as

$$y' = \frac{dy}{dx} \quad \text{and} \quad y'' = \frac{d^2y}{dx^2}$$

Then Equation 7.64 can be written in shorthand notation as

$$a(x)y'' + b(x)y' + c(x)y = f(x) \tag{7.65}$$

If the right-hand side of Equation 7.65 is identically zero (i.e., $f(x) = 0$), then

$$a(x)y'' + b(x)y' + c(x)y = 0 \tag{7.66}$$

This is called the *homogeneous equation* as it only contains terms in y and its derivatives.

Linear constant coefficient second-order differential equations are a special case of Equation 7.65 where $a(x)$, $b(x)$ and $c(x)$ are all constants (i.e., only numbers). So, the differential equations to be solved in this section are of the following type:

$$ay'' + by' + cy = f(x) \tag{7.67}$$

where a, b, and c are constants.

The general solution of Equation 7.67 is made up of two parts:

General solution = Complementary function + Particular integral
$$y(t) = y_{CF} \quad + \quad y_{PI} \tag{7.68}$$

The task is to find these two functions y_{CF} and y_{PI} separately and then add them together to get the final solution. The next section considers how to compute the complementary function y_{CF}.

7.4.1   Complementary Function (CF)

The complementary function is obtained by solving Equation 7.67 with $f(x) = 0$, that is, the homogenous equation:

$$ay'' + by' + cy = 0 \tag{7.69}$$

Note: A second-order linear differential equation always has two and only two linearly independent complementary solutions.

7.4.1.1   General Solution for the Complementary Function

If $y_1(x)$ and $y_2(x)$ are two linearly independent solutions to the homogeneous equations:

$$ay'' + by' + cy = 0 \tag{7.70}$$

Then $y_1(x)$ and $y_2(x)$ are both solutions to Equation 7.70, that is,

$$ay_1'' + by_1' + cy_1 = 0 \tag{7.71}$$

$$ay_2'' + by_2' + cy_2 = 0 \tag{7.72}$$

Then the linear combination of $y_1(x)$ and $y_2(x)$

$$y_{CF}(t) = \alpha y_1(x) + \beta y_2(x) \qquad \text{(Principle of superposition)} \tag{7.73}$$

where α and β are constants, is also a solution to Equation 7.70.

Note: Equation 7.73 is called the general solution for the complementary function.

7.4.1.2 How to Find the Complementary Function

The complementary function is given as the solution to the homogeneous Equation 7.70, that is, with $f(x) = 0$:

$$ay'' + by' + cy = 0 \tag{7.74}$$

It is obtained by assuming that Equation 7.70 has solutions of the exponential form

$$y = e^{mx} \tag{7.75}$$

where m is a constant (real or complex) that needs to be determined.

For $y = e^{mx}$ to be a solution to Equation 7.70, it must satisfy Equation 7.70. Starting with y and calculating its first and second derivatives y' and y'' gives

$$y' = me^{mx}, \qquad y'' = m^2 e^{mx} \tag{7.76}$$

Substituting y, y', and y'' back into Equation 7.70 gives

$$am^2 e^{mx} + bme^{mx} + ce^{mx} = 0 \tag{7.77}$$

$$e^{mx}(am^2 + bm + c) = 0 \tag{7.78}$$

Since e^{mx} does not equal zero implies for Equation 7.78 to be equal to zero, then the quadratic equation must be equal to zero, that is,

$$am^2 + bm + c = 0 \tag{7.79}$$

Equation 7.79 is given a special name called the *characteristic* or *auxiliary equation*.

Next, an example is shown of how to calculate the complementary function y_{CF}.

Example 7.15

Solve the following second-order linear constant coefficient differential equation:

$$y'' + 3y' + 2y = 0 \tag{7.80}$$

Solution: Assume solution of the form $y = e^{mx}$, then $y' = me^{mx}$ and $y'' = m^2 e^{mx}$.

Substituting for y, y', and y'' back into Equation 7.80 gives

$$m^2 e^{mx} + 3me^{mx} + 2e^{mx} = 0 \tag{7.81}$$

$$e^{mx}(m^2 + 3m + 2) = 0 \tag{7.82}$$

Now, the term $e^{mx} \neq 0$.

This means that the quadratic equation must be equal to zero, that is,

$$m^2 + 3m + 2 = 0 \tag{7.83}$$

This factorizes as $(m + 2)(m + 1) = 0$, which give the values for m as $m = -2$, $m = -1$. Substituting these values of m back into $y = e^{mx}$, gives the two linearly independent solutions as

$$y_1 = e^{-2x} \text{ and } y_2 = e^{-x} \tag{7.84}$$

The general complementary solution to Equation 7.80 is now given as a general combination of y_1 and y_2, that is,

$$y = \alpha e^{-2x} + \beta e^{-x} \tag{7.85}$$

where α and β are arbitrary constants to be determined from initial conditions.

7.4.2   Types of Solutions

It has already been shown that by assuming an exponential solution of the form $y = e^{mx}$ to the constant coefficient equation $ay'' + by' + cy = 0$ gives rise to the characteristic or auxiliary equation $am^2 + bm + c = 0$, which is solved for m in order to fully determine the general complementary solution to the equation $ay'' + by' + cy = 0$.

The general solution to the characteristic equation $am^2 + bm + c = 0$ is given by

$$m = \frac{-b \pm \sqrt{b^2 - 4ac}}{2a} \tag{7.86}$$

and the roots depend on the sign of the discriminant $\Delta = b^2 - 4ac$.

There are three cases to consider:

7.4.2.1   Case 1: Real and Distinct Roots m_1 and m_2

If $b^2 - 4ac > 0$, then the solutions of the characteristic Equation 7.79 are real and different as follows:

$$m = m_1 \text{ and } m = m_2$$

where

$$m_1 = \frac{-b + \sqrt{b^2 - 4ac}}{2a} \text{ and } m_1 = \frac{-b - \sqrt{b^2 - 4ac}}{2a}$$

The general solution to the complementary function is given as

$$y_{CF} = \alpha e^{m_1 x} + \beta e^{m_2 x} \tag{7.87}$$

Example 7.16

Solve

$$y'' + 5y' + 6y = 0 \tag{7.88}$$

Solution: Assume the answer in the form $y = e^{mx}$.
The characteristic equation is

$$m^2 + 5m + 6 = 0$$

$$\Rightarrow (m + 2)(m + 3) = 0$$

$$m = -2 \text{ and } m = -3$$

$$y_{CF} = \alpha e^{-2x} + \beta e^{-3x}$$

7.4.2.2   Case 2: Real and Repeated Roots

If $b^2 - 4ac = 0$, then the solutions of the characteristic Equation 7.79 are repeated roots given as

$$m_1 = -\frac{b}{2a}$$

There is one linearly independent solution, which is given by $y_1 = e^{m_1 x}$. To construct the general solution to the complementary function, there needs to be a second linearly independent solution.

It can be shown that if $y_1 = e^{m_1 x}$ is one solution to Equation 7.74, then the second independent solution is $y_2 = xe^{m_1 x}$.

Note: If one solution was $q_1 = e^{m_1 t}$, then the other would be $q_2 = te^{m_1 t}$.

The general solution for the complementary function is now given as

$$y_{CF} = \alpha e^{m_1 x} + \beta x e^{m_1 x} \tag{7.89}$$

Example 7.17

Solve

$$y'' + 6y' + 9y = 0 \qquad (7.90)$$

Solution: Assume the answer is of the form $y = e^{mx}$.

The characteristic equation is $m^2 + 6m + 9 = 0$, which implies $(m + 3)(m + 3) = 0$, $m = -3$ is a repeated root. So one solution is $y_1 = e^{-3x}$ and the second solution is $y_2 = xe^{-3x}$.

The general solution for the complementary function is now

$$y_{CF} = \alpha e^{-3x} + \beta x e^{-3x}$$

which can also be written as $y_{CF} = (\alpha + \beta x)e^{-3x}$ by taking e^{-3x} as a common factor.

7.4.2.3 Case 3: Complex Conjugate Roots

If $b^2 - 4ac < 0$, then the solutions to the characteristic equation are $m = m_1$ and $m = m_2$, where m_1 and m_2 are complex conjugate solutions.

$$m_1 = p + jq \quad \text{and} \quad m_2 = p - jq$$

with $p = -\dfrac{b}{2a}$ and $q = \dfrac{\sqrt{4ac - b^2}}{2a}$.

Then the solution to the complementary function is

$$y_{CF} = \alpha e^{(p+jq)x} + \beta e^{(p-jq)x} \qquad (7.91)$$

Taking out the e^{px} as a common factor gives

$$y_{CF} = e^{px}\left(\alpha e^{jqx} + \beta e^{-jqx}\right) \qquad (7.92)$$

Since this contains imaginary exponentials, these can be replaced by using Euler's identity:

$$e^{j\theta} = \cos\theta + j\sin\theta$$

This then gives the following:

$$e^{jqx} = \cos qx + j\sin qx \quad \text{and} \quad e^{-jqx} = \cos qx - j\sin qx$$

Therefore, the general solution for the complementary function can now be written as

$$y_{CF} = e^{px}(A\cos qx + B\sin qx) \qquad (7.93)$$

where A and B are constants.

Example 7.18

Solve

$$y'' + 4y' + 9y = 0 \qquad (7.94)$$

Solution: Assume the answer in the form $y = e^{mx}$.

The characteristic equation is $m^2 + 4m + 9 = 0$. This can be solved using the quadratic formula to give the solutions as a complex conjugate pair:

$$m = -2 \pm j\sqrt{5}$$

So, here $p = -2$ and $q = \sqrt{5}$, giving the general solution to the complementary function as

$$y_{CF} = e^{-2x}(A\cos\sqrt{5}x + B\sin\sqrt{5}x) \qquad (7.95)$$

7.4.3   Particular Integral (P.I.)

So far, consideration has been given to second-order differential equations of the form

$$ay'' + by' + cy = 0 \qquad (7.96)$$

that is, where $f(x) = 0$ has been made. Now considering the full differential equation:

$$ay'' + by' + cy = f(x) \qquad (7.97)$$

Substituting in the complementary function would make the left-hand side of Equation 7.97 equal to zero and not $f(x)$, so there must be a further term in the solution that will make the left-hand side equal to $f(x)$ and not equal to zero.

This extra function is called the particular integral (P.I.). If the solutions to the complementary function are real and different solutions, let the particular integral be called $X(x)$. Then the complete solution $y(x)$ will be of the form

$$y = \alpha e^{m_1 x} + \beta x e^{m_1 x} + X(x) \qquad (7.98)$$

$X(x)$ is an extra function known as the particular integral (yet to be determined).

7.4.3.1   How to Find the Particular Integral

To find the particular integral, one assumes the most general form of the function on the right-hand side of Equation 7.97.

$$ay'' + by' + cy = f(x)$$

that is, general form of the function $f(x)$,

Table 7.1 Particular Integral y_{PI} for Different Functions $f(x)$

$f(x)$	Try y_{PI} as
k (constant)	$y_{PI} = C$
kx	$y_{PI} = Cx + D$
kx^2	$y_{PI} = Cx^2 + Dx + E$
$k \sin wx$ or $k \cos wx$	$y_{PI} = C \sin wx + D \cos wx$
αe^{kx}	$y_{PI} = Ce^{kx}$

Table 7.1 shows which functions for the particular integral to try for the different functions of $f(x)$.

Note: The constants C, D, E, and so on are determined by substituting y_{PI} and its derivatives back into Equation 7.97 and equating coefficients on both sides.

Example 7.19

If $f(x) = \sin 4x$, then try the particular integral as $y_{PI} = C \sin 4x + D \cos 4x$.

If $f(x) = f(x) = x + 2e^x$, then try the particular integral as $y_{PI} = Cx + D + Ee^x$.

Finally, putting all the theory together for the complementary function and the particular integral, the next example shows how to solve a general second-order differential equation.

Example 7.20

Solve

$$y'' - 5y' + 6y = 2\sin 4x \qquad (7.99)$$

subject to initial conditions, $x = 0$, $y = \dfrac{27}{25}$, and $y' = \dfrac{117}{50}$.

Solution: The general solution is given by $y(t) = y_{CF} + y_{PI}$.

First Finding the Complementary Function (y_{CF})

Making the right-hand side of Equation 7.99 equal to zero, $f(x) = 0$, gives

$$y'' - 5y' + 6y = 0 \qquad (7.100)$$

Try a solution of the form $y = e^{mx}$.
The characteristic equation is

$$m^2 - 5m + 6 = 0$$

$$\Rightarrow (m - 2)(m - 3) = 0$$

$m = 2$ and $m = 3$ (this is case 1 for different types of solutions) implies the solution is

$$y_{CF} = \alpha e^{2x} + \beta e^{3x}$$

For the Particular Integral (y_{PI})

Look at the right-hand side of Equation 7.99. The function is $f(x) = 2 \sin 4x$.
Try

$$y_{PI} = A \cos 4x + B \sin 4x \qquad (7.101)$$

Differentiating Equation 7.101 twice gives

$$y'_{PI} = -4A \sin 4x + 4B \cos 4x$$

$$y''_{PI} = -16A \cos 4x - 16B \sin 4x$$

Now by substituting y_{PI}, y_{PI}' and y_{PI}'' back into Equation 7.99 gives

$$-16A \cos 4x - 16B \sin 4x - 5(-4A \sin 4x + 4B \sin 4x) + 6(A \cos 4x + B \sin 4x) = 2 \sin 4x$$

Collecting all the cos $4x$ and sin $4x$ terms together on the left-hand side gives

$$(-16A + 6A - 20B) \cos 4x + (-16B + 20A + 6B) \sin 4x = 2 \sin 4x + 0 \cos 4x$$

$$(-10A - 20B) \cos 4x + (-10B + 20A) \sin 4x = 2 \sin 4x + 0 \cos 4x$$

Here, the right-hand side has a 0 multiplying the cos $4x$, so like terms can be compared on both sides.

By equating the coefficients of the cosine terms on both sides gives

$$-10A - 20B = 0 \quad \Rightarrow \quad -10A = 20B$$

$$\Rightarrow \quad A = -2B \qquad (7.102)$$

By equating coefficients of the sine terms on both sides gives,

$$-10B + 20A = 2 \quad \Rightarrow \quad -10B = 2 - 20A$$

$$\Rightarrow \quad -5B = 1 - 10A \qquad (7.103)$$

Solving Equations 7.102 and 7.103 simultaneously for A and B gives

$$A = \frac{2}{25} \quad \text{and} \quad B = -\frac{1}{25} \qquad (7.104)$$

Therefore the particular integral can be obtained by using these values of A and B back into Equation 7.101 to give

$$y_{PI} = \frac{2}{25} \cos 4x - \frac{1}{25} \sin 4x = \frac{1}{25}(2 \cos 4x - \sin 4x) \qquad (7.105)$$

The general solution is given by

$$y(t) = y_{CF} + y_{PI}$$

$$y(t) = \alpha e^{2x} + \beta e^{3x} + \frac{1}{25}(2\cos 4x - \sin 4x) \qquad (7.106)$$

Now to finally finish off the complete solution, the constants α and β need to be found using the given initial conditions. The initial conditions are $x = 0, y = \frac{27}{25}$, and $y' = \frac{117}{50}$.

Using $x = 0$ and $y = \frac{27}{25}$ in Equation 7.106 gives

$$\alpha + \beta = 1 \qquad (7.107)$$

Differentiating Equation 7.106 gives

$$y'(t) = 2\alpha e^{2x} + 3\beta e^{3x} - \frac{1}{25}(8\cos 4x + 4\sin 4x) \qquad (7.108)$$

Using $x = 0$ and $y' = \frac{117}{50}$ in Equation 7.108 gives

$$2\alpha + 3\beta = 2.5 \qquad (7.109)$$

Solving Equations 7.107 and 7.109 simultaneously gives

$$\alpha = 0.5 \text{ and } \beta = 0.5 \qquad (7.110)$$

The complete solution to the original differential Equation 7.99 is given as

$$y(t) = 0.5e^{2x} + 0.5e^{3x} + \frac{1}{25}(2\cos 4x - \sin 4x) \qquad (7.111)$$

The next section is of applications that show how differential equations arise in real-world engineering problems and how using the different methods studied so far can help solve these real problems.

7.5 Applications

Example 7.21: Calculating the Time for the Smoke Layer to Develop

In this problem, as the fire progresses, the dangerous smoke layer develops as shown in Figure 7.2. The time for the smoke layer to reach a certain height can be calculated by solving a differential equation. In Figure 7.2, room height is H and lower layer height is z. The fire is treated as a point source of heat $\dot{Q}$. The mass flow rate of the lower layer to the upper layer is given by $\dot{m}_p$, the plume mass flow rate. The plume is considered only as a means of transporting mass from the lower layer to the upper layer.

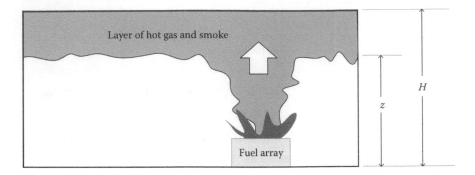

Figure 7.2 Two-zone model for the smoke layer.

Applying the conservation of mass and energy to the lower layer gives a simplified differential equation for smoke filling (in dimensionless form) as follows:

$$\frac{dy}{d\tau} + 0.21\left(\dot{Q}*\right)^{\frac{1}{3}} y^{\frac{5}{3}} = 0 \qquad (7.112)$$

given $y = 1$ when $\tau = 0$.

The dimensionless height given by $y = \dfrac{z}{H}$ varies from 0 to 1 and gives the fraction of room height below the smoke layer. The dimensionless time is given by $\tau = t\dfrac{H^2}{S}\sqrt{\dfrac{g}{H}}$, where t is time in seconds, g equals 9.81 ms^{-2}, H is the room height, and S the floor area.

Equation 7.112 is an example of a differential equation that can be solved using the method of separating variables as follows.

Subtracting $0.21\left(\dot{Q}*\right)^{\frac{1}{3}} y^{\frac{5}{3}}$ from both sides of Equation 7.112 gives,

$$\frac{dy}{d\tau} = -0.21\left(\dot{Q}*\right)^{\frac{1}{3}} y^{\frac{5}{3}} \qquad (7.113)$$

Dividing by $y^{\frac{5}{3}}$ and integrating both sides of Equation 7.113 with respect to $d\tau$ gives

$$\int y^{-\frac{5}{3}} dy = \int -0.21\left(\dot{Q}*\right)^{\frac{1}{3}} d\tau \qquad (7.114)$$

$$-\frac{3}{2} y^{-\frac{2}{3}} = -0.21\left(\dot{Q}*\right)^{\frac{1}{3}} \tau + C \qquad (7.115)$$

using the initial conditions $y = 1$ when $\tau = 0$ gives $C = -\dfrac{3}{2}$.

$$-\frac{3}{2} y^{-\frac{2}{3}} = -0.21\left(\dot{Q}*\right)^{\frac{1}{3}} \tau - \frac{3}{2} \qquad (7.116)$$

$$y^{-\frac{2}{3}} = 0.14\left(\dot{Q}*\right)^{\frac{1}{3}} \tau + 1 \qquad (7.117)$$

So now we have an equation relating τ and y as follows:

$$0.14\left(\dot{Q}*\right)^{\frac{1}{3}} \tau = y^{-\frac{2}{3}} - 1 \qquad (7.118)$$

This Equation 7.118 can now be used to calculate the time t taken for a room to fill with smoke to any height.

If a pool of kerosene is ignited releasing 186 kW in a room with floor area 5.62 m by 5.62 m and a height of 5.95 m, the time until the smoke layer has filled half the room can be calculated as follows: First, calculate $\dot{Q}*$ using the following formula:

$$\dot{Q}* = \frac{\dot{Q}}{1100H^{\frac{5}{2}}} = \frac{186}{1100(5.95)^{\frac{5}{2}}} = 0.002 \qquad (7.119)$$

For half the room, $y = 0.5$, and so Equation 7.118 now gives τ as

$$0.14(0.002)^{\frac{1}{3}} \tau = (0.5)^{-\frac{2}{3}} - 1$$

Rearranging this to find τ yields $\tau = 33.3$ seconds.
Now since

$$\tau = t \frac{H^2}{S} \sqrt{\frac{g}{H}} = 1.44t$$

this implies that the time is $t \approx 23$ seconds. It therefore takes less than half a minute for a 186 kW fire to fill half the room with smoke.

Note: A 186 kW fire is approximately a wastepaper bin sized fire.

Example 7.22: Probabilistic Modeling Using Continuous Markov Processes

Using a probabilistic model of system behavior for a continuous time Markov process, the resulting model can be described by a set of differential equations known as the Kolmogorov forward equations. Consider a simple two-state problem in which a system is either working (R) or not working (F), as shown in Figure 7.3.

For a two-state dynamical Markov process represented by Figure 7.3, λ is the failure rate and μ is the repair rate, which are both constants.

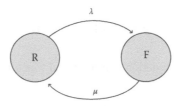

Figure 7.3 Two-state dynamical process. The R state represents the system working and the F state represents not working.

The rate of change of the probability of being in state R is decreased by λ and increased by μ like a probability flow and are given by the Kolmogorov forward equations:

$$\frac{dR}{dt} = -\lambda R + \mu F \tag{7.120}$$

where λR is the probability of the system operating at time t and not failing in the time step dt, and μF is the probability of the system being in the failed state at time t and being repaired in the time step dt.

Similarly, for the rate of change of the system being in state F is given by

$$\frac{dF}{dt} = -\mu F + \lambda R \tag{7.121}$$

Equations 7.120 and 7.121 are first-order differential equations that need to be solved to find the probabilities of being in each state.

Solution: Starting with the equation for the R state (Equation 7.120) and having the initial condition that $R(0) = R_0$ (i.e., the initial probability of the system working)

$$\frac{dR}{dt} = -\lambda R + \mu F \tag{7.122}$$

Replacing for $F = 1 - R$, into Equation 7.121 gives

$$\frac{dR}{dt} = -\lambda R + \mu(1 - R) = -\lambda R + \mu - \mu R \tag{7.123}$$

This tidies up as

$$\frac{dR}{dt} = \mu - (\lambda + \mu)R \tag{7.124}$$

which can be written as

$$\frac{dR}{dt} + (\lambda + \mu)R = \mu \text{ with initial condition } R(0) = R_0 \tag{7.125}$$

Equation 7.125 is an example of a differential equation that can be solved using the integration factor method.

Comparing Equation 7.125 with the standard form equation

$$\frac{dy}{dt} + P(t)y = Q(t) \tag{7.126}$$

gives $P(t) = (\lambda + \mu)$ and $Q(t) = \mu$.

The integrating factor (I.F.) then becomes

$$\text{I.F.} = e^{\int (\lambda + \mu) dt} = e^{(\lambda + \mu)t} \tag{7.127}$$

The solution is given by Equation 7.47 as

$$R(t)e^{(\lambda + \mu)t} = \int \mu e^{(\lambda + \mu)t} \, dt \tag{7.128}$$

$$R(t)e^{(\lambda + \mu)t} = \frac{\mu}{(\lambda + \mu)} e^{(\lambda + \mu)t} + C \tag{7.129}$$

$$R(t) = \frac{\mu}{(\lambda + \mu)} + Ce^{-(\lambda + \mu)t} \tag{7.130}$$

Using the initial condition $t = 0$, $R = R_0$ gives

$$C = \left(R_0 - \frac{\mu}{(\lambda + \mu)} \right)$$

So finally, the solution for $R(t)$ is

$$R(t) = \frac{\mu}{(\lambda + \mu)} + \left(R_0 - \frac{\mu}{(\lambda + \mu)} \right) e^{-(\lambda + \mu)t} \tag{7.131}$$

A similar relationship can be found for $F(t)$ using $F(t) = 1 - R(t)$ as

$$F(t) = \frac{\lambda}{(\lambda + \mu)} + \left(F_0 - \frac{\lambda}{(\lambda + \mu)} \right) e^{-(\lambda + \mu)t} \tag{7.132}$$

These solutions then give the probabilities for the system to be in a working or not working state.

Example 7.23: Finding the Current in a *RL* Circuit

Consider the practical electrical *RL* circuit shown in Figure 7.4, The problem is to determine the current $i(t)$ in the circuit.

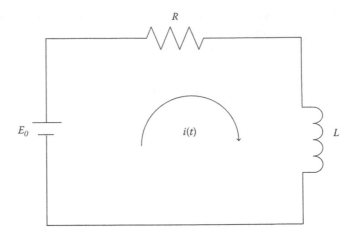

Figure 7.4 An electrical *RL* series circuit.

Using Kirchhoff's voltage law gives the differential equation for the circuit as

$$L\frac{di}{dt} + Ri = E_0 \qquad (7.133)$$

subject to the initial conditions $i = 0$ when $t = 0$.

Notes:
- *The values of L the inductance, R the resistance, and E_0 the voltage source are constants.*
- *Equation 7.133 is a type of differential equation in which there is a choice of methods that can be used to solve it, either by separating variables or the integrating factor method. The solution shows the method of separating variables.*

Solution: Starting with Equation 7.133 and subtracting the term Ri from both sides yields

$$L\frac{di}{dt} = E_0 - Ri \qquad (7.134)$$

Dividing both sides of Equation 7.134 by $(E_0 - Ri)$ gives

$$\frac{L}{(E_0 - Ri)}\frac{di}{dt} = 1 \qquad (7.135)$$

Integrating both sides of Equation 7.135 with respect to dt gives

$$\int\left(\frac{L}{E_0 - Ri}\right)dt = \int 1\, dt \qquad (7.136)$$

Integrating both sides of Equation 7.136 gives

$$-\frac{L}{R}\ln(E_0 - Ri) = t + C \tag{7.137}$$

Now using the initial conditions $i = 0$ when $t = 0$ gives C as

$$C = -\frac{L}{R}\ln E_0 \tag{7.138}$$

Substituting this value of C back into Equation 7.137 and rearranging gives

$$\ln\left(\frac{E_0}{E_0 - Ri}\right) = \frac{R}{L}t \tag{7.139}$$

Now to find $i(t)$, Equation 7.139 is raised to the power e giving

$$\frac{E_0}{E_0 - Ri} = e^{\frac{R}{L}t} \tag{7.140}$$

Inverting both sides of Equation 7.140 gives

$$\frac{E_0 - Ri}{E_0} = e^{-\frac{R}{L}t} \tag{7.141}$$

Rearranging this to make $i(t)$ the subject gives the final solution for the current in the circuit as

$$i(t) = \frac{E_0}{R}\left(1 - e^{-\frac{R}{L}t}\right) \tag{7.142}$$

Problems

Solve the following differential equations using the appropriate method.

7.1 $\dfrac{dy}{dx} = \dfrac{x}{y}$ $\qquad\qquad$ with $y(0) = 1$

7.2 $\dfrac{dy}{dx} = (y - 3)(x + 5)$

7.3 $\dfrac{dy}{dx} = \dfrac{1}{xy + x}$

7.4 $x\dfrac{dy}{dx} - y = x^3$

7.5 $(1 - x^2)\dfrac{dy}{dx} - 2xy = x^4$ with $y(0) = 1$

7.6 $\dfrac{d^2y}{dx^2} - \dfrac{dy}{dx} - 2y = x + 2$

7.7 $\dfrac{d^2y}{dx^2} - 10\dfrac{dy}{dx} + 25y = 10$

7.8 $\dfrac{d^2y}{dx^2} + 4\dfrac{dy}{dx} + 5y = 13e^{3x}$ with $y(0) = 0$ and $y'(0) = 1.5$

7.9 $\dfrac{dT}{dt} = 1 + 5T$ with $T(0) = 1$

7.10 Consider the following RC circuit in Figure 7.5. Kirchhoff's voltage law for the circuit gives the following equation.

$$Ri + \frac{1}{C}\int i\,dt = E_0$$

where R, C, and E_0 are all constants. By differentiating the above equation, derive a differential equation for the current $i(t)$ in the circuit and hence show that $i(t) = ke^{-\frac{t}{RC}}$, where k is a constant.

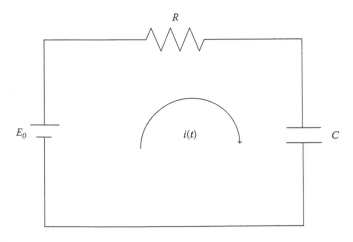

Figure 7.5 An electrical *RC* series circuit.

8 Laplace Transforms

The Laplace Transform is an integral transform named after its founder Pierre-Simon Laplace. It takes a function of continuous time t ($t > 0$) to a function of a complex variable s (frequency). As discussed earlier in Chapter 7, when modeling real world problems, the formulation of differential equations naturally arises in many different fields of engineering.

8.1 Why Do We Need the Laplace Transform?

The Laplace transform is a very important tool in engineering disciplines as it enables the following:

1. It helps to solve linear differential equations with given initial conditions, for systems that can be described by the following types of equations:

$$a\frac{d^2y}{dx^2} + b\frac{dy}{dx} + cy = E \tag{8.1}$$

2. The method is also particularly useful if the inputs to the differential equations, that is, the E term in Equation 8.1 are discontinuous inputs like the unit step function.

3. Incorporates the initial conditions at the start of the solution to the problem.

4. In systems engineering, the system is broken down into components as blocks. Each block can be represented in the s-domain and then manipulated.

8.2 Derivation from a Power Series

Most mathematics textbooks will start with the formula for the Laplace transform without any reference to how it comes about mathematically. Here, it is more appropriate to first consider how the Laplace transform is derived.

Starting with a *discrete power series*, this can be written as follows:

$$\sum_{0}^{\infty} a_n x^n = a_0 + a_1 x + a_2 x^2 + \ldots$$

And for some a_n this series can be written in closed form as, say, $A(x)$:

$$\sum_{0}^{\infty} a_n x^n = A(x)$$

Using a slightly different notation for the coefficients $a_n = a(n)$ this becomes

$$\sum_{0}^{\infty} a(n) x^n = A(x) \tag{8.2}$$

So, different values of $a(n)$ can produce a different closed form sum $A(x)$. Some examples of using specific $a(n)$'s are given next.

Example 8.1

If $a(n) = 1$, then the series just becomes

$$\sum_{0}^{\infty} x^n = 1 + x + x^2 + \ldots$$

This is just the geometric series with first term $a = 1$ and common ratio $r = x$.

This has a sum to infinity given by

$$S_\infty = \frac{a}{1 - r}$$

So, the series in closed form becomes

$$A(x) = \frac{1}{1 - x}$$

provided $|x| < 1$

Example 8.2

If $a(n) = \dfrac{1}{n!}$, then the series is

$$\sum_{0}^{\infty} \frac{x^n}{n!} = 1 + \frac{x}{1!} + \frac{x^2}{2!} + \dots$$

This is just the series for the exponential function of x. Therefore, $A(x) = e^x$ in closed form.

So far, the power series has been one using a discrete summation. Now considering the *continuous analog* of the above case one obtains the corresponding Laplace transform as follows.

Replace the summation $\displaystyle\sum_{0}^{\infty}$ with the integral $\displaystyle\int_{0}^{\infty}$ for continuous time. Replace the discrete integers $n = 0, 1, 2, 3, \dots$ by continuous time t such that t goes between $0 < t < \infty$ for all values of t. Therefore, Equation 8.2 becomes

$$\int_{0}^{\infty} a(t)x^t \ dt = A(x) \tag{8.3}$$

Equation 8.3 could be left in this form, but for integration purposes it is not good to have x as the base. It is usually better to have e as the function.

This can be done by letting $x = e^{\ln(x)}$, so $x^t = (e^{\ln(t)})^t$

Now for this integral to converge, as $t \to \infty$, x has be less than 1 and $x > 0$ to avoid any imaginary numbers appearing. So, $0 < x < 1$ implies that $\ln(x) < 0$ in this range for x.

Let $-s = \ln(x)$ and replace $a(t) = f(t)$ in Equation 8.3:

$$\int_{0}^{\infty} f(t)e^{-st} \ dt = F(s) \tag{8.4}$$

This is called the *Laplace transform* of the function $f(t)$. This is just the *continuous analog of the summation of a power series*.

8.3   Introduction and Standard Transforms

If $f(t)$ is a piecewise continuous function for $t \geq 0$, then the Laplace transform of $f(t)$ is defined as

$$\mathcal{L}\big[f(t)\big] = \int_{0}^{\infty} f(t)e^{-st} \ dt = F(s) \tag{8.5}$$

Note: $s = \sigma + jw$ (that is in general a complex variable).

The Laplace transform pair can be denoted as follows:

$$
f(t) \quad \underset{\substack{\leftarrow \\ \mathcal{L}^{-1}}}{\overset{\substack{\mathcal{L} \\ \rightarrow}}{}} \quad F(s)
$$

t-domain $\qquad$ s-domain

The operator $\mathcal{L}$ is used to represent the Laplace transform and the operator $\mathcal{L}^{-1}$ is used to represent the inverse Laplace transform.

8.3.1 Schematic Representation of Laplace Transforms

The Laplace transform method for solving differential equations can be more easily understood using the schematic diagram in Figure 8.1 showing the process being applied. From the diagram, it can be seen that the differential equation to be solved is first Laplace transformed. Second, inputting the initial conditions then results in an algebraic expression for the variable in the s-domain. Since this part is all algebraic, it is fairly easy to manipulate to determine an expression for $Y(s)$. Finally, use the inverse Laplace transforms to generate the solution back in the time domain, $y(t)$, and hence, obtaining the solution to the original problem.

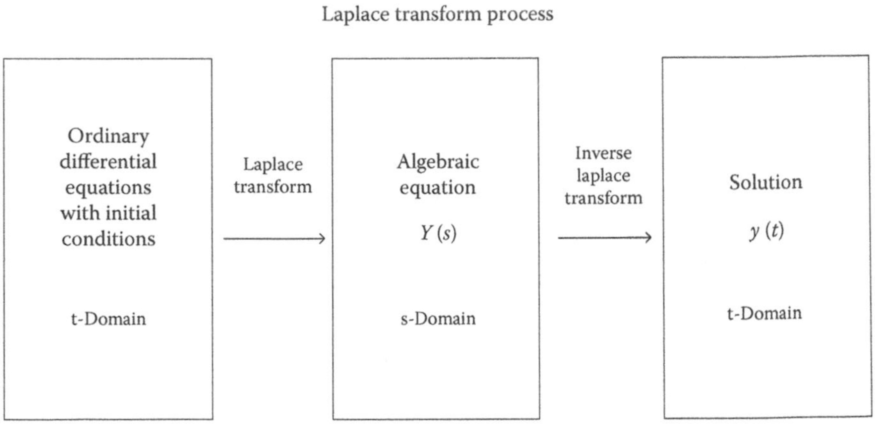

Laplace transform process

Ordinary differential equations with initial conditions t-Domain	Laplace transform $\longrightarrow$	Algebraic equation $Y(s)$ s-Domain	Inverse laplace transform $\longrightarrow$	Solution $y(t)$ t-Domain

Figure 8.1 Schematic representation of the Laplace transform process.

8.3.2 Standard Transforms

In the aforementioned process, the differential equation needs to be Laplace transformed and so this requires Laplace transforming different functions of time as well as derivative terms like $\dfrac{dy}{dt}$ and $\dfrac{d^2 y}{dt^2}$ etc.

Most of the Laplace transforms and subsequently the inverse Laplace transforms are generally taken from a standard table of results. Therefore, it is useful to first show some of the more common Laplace Transforms of functions in a table format and then to see how they are derived from first principles using the formula definition. Table 8.1 gives some functions and their corresponding Laplace transforms. All the functions in the table can be proved using the basic definition given by Equation 8.5 and some further manipulation. Next, a few of the results are proved to show how they are derived using the formula definition.

Table 8.1 Standard Laplace Transforms of Some Common Functions

$f(t)$	$F(s)$
1	$\dfrac{1}{s}$
t	$\dfrac{1}{s^2}$
e^{at}	$\dfrac{1}{s-a}$
$\sin(at)$	$\dfrac{a}{s^2+a^2}$
$\cos(at)$	$\dfrac{s}{s^2+a^2}$
t^n	$\dfrac{n!}{s^{n+1}}$
$t^n\,e^{-at}$	$\dfrac{n!}{(s+a)^{n+1}}$

Example 8.3

Laplace transform $f(t) = 1$.

If $f(t) = 1$, then using formula $F(s) = \int_0^\infty f(t)e^{-st}\,dt$ gives

$$= \int_0^\infty 1\,e^{-st}\,dt = \left[\frac{e^{-st}}{-s}\right]_0^\infty = (0) - \left(\frac{1}{-s}\right) = \frac{1}{s}$$

$$F(s) = \frac{1}{s}$$

Example 8.4

Laplace transform $f(t) = t$.

If $f(t) = t$, then $F(s) = \int_0^\infty t\,e^{-st}\,dt$. Here use integration by parts (Chapter 6, Section 6.3.3), that is, using the formula

$$\int u.\frac{dv}{dt}\,dt = u.v - \int v.\frac{du}{dt}\,dt$$

Let $u = t$ $\quad \dfrac{dv}{dt} = e^{-st}$ $\quad$ gives $\quad \dfrac{du}{dt} = 1$ $\quad$ and $\quad v = \dfrac{e^{-st}}{-s}$

This gives

$$F(s) = \left[t\frac{e^{-st}}{-s}\right]_0^\infty - \int_0^\infty \frac{e^{-st}}{-s}\,dt = 0 + \int_0^\infty \frac{e^{-st}}{s}\,dt = \left[\frac{e^{-st}}{-s^2}\right]_0^\infty = \frac{1}{s^2}$$

$$F(s) = \frac{1}{s^2}$$

Example 8.5

Laplace transform $f(t) = e^{at}$.

If $f(t) = e^{at}$, then $F(s) = \int\limits_0^\infty e^{at} e^{-st}\, dt$

$$F(s) = \int\limits_0^\infty e^{-t(s-a)}\, dt = \left[\frac{e^{-t(s-a)}}{-(s-a)} \right]_0^\infty = 0 - \frac{1}{-(s-a)}$$

$$F(s) = \frac{1}{(s-a)}$$

Note: Other Laplace transforms such as $\sin at$ and $\cos at$ can be done without the need of integration by parts in a much more convenient manner using Euler's formula and complex numbers.

Euler's formula is $e^{j\theta} = \cos\theta + j\sin\theta$, so it implies that $e^{jat} = \cos at + j\sin at$. Since this formula contains a real part $\cos at$ and imaginary part $\sin at$, then if one Laplace transforms e^{jat} the real part will be the Laplace transform of $\cos at$ while the imaginary part will be the Laplace transform of $\sin at$. This is shown in Example 8.6.

Example 8.6

$f(t) = e^{jat}$, then $F(s) = \int\limits_0^\infty e^{jat} e^{-st}\, dt$

$$F(s) = \int\limits_0^\infty e^{-(s-ja)t}\, dt$$

Therefore,

$$F(s) = \left[\frac{e^{-(s-ja)t}}{-(s-ja)} \right]_0^\infty$$

$$F(s) = \left[0 - \left(-\frac{1}{s-ja} \right) \right]$$

$$F(s) = \frac{1}{s-ja},$$

Now this needs to be written as a real and imaginary part, that is, $a + jb$. One can multiply the top and bottom by the complex conjugate of the bottom $(s - ja)$, that is, by $(s + ja)$, to give

$$F(s) = \frac{(s+ja)}{(s-ja)(s+ja)} = \frac{s+ja}{s^2+a^2} = \frac{s}{s^2+a^2} + j\frac{a}{s^2+a^2}$$

So this gives

$$\mathcal{L}\left[e^{jat}\right] = \mathcal{L}\left[\cos at + j\sin at\,\right] = \frac{s}{s^2 + a^2} + j\frac{a}{s^2 + a^2}$$

from which it can be seen that

$$\mathcal{L}\left[\cos at\right] = \frac{s}{s^2 + a^2}$$

and

$$\mathcal{L}\left[\sin at\right] = \frac{a}{s^2 + a^2}$$

as required.

8.3.3 Linearity of Laplace Transforms

From the definition of integration, it follows that if we have two functions $f(t)$ and $g(t)$, and these are both of exponential order, then

$$\mathcal{L}\left[af(t) + bg(t)\right] = a\mathcal{L}\left[f(t)\right] + b\mathcal{L}\left[g(t)\right] = aF(s) + bG(s) \qquad (8.6)$$

This follows from the property that the integral of the sum of two functions is equal to the sum of the two separate integrals. This property of *linearity* allows the Laplace transforms of sums of functions to be found easily. The next example shows how this is can be applied.

Example 8.7

Given $f(t) = t$ and $g(t) = e^{3t}$, then using Equation 8.6 gives

$$\mathcal{L}\left[5t - 11e^{3t}\right] = 5\mathcal{L}[t] - 11\mathcal{L}\left[e^{3t}\right] = \frac{5}{s^2} - \frac{11}{s-3}$$

and, conversely, it follows for the inverse Laplace transforms:

$$5t - 11e^{3t} = \mathcal{L}^{-1}\left[\frac{5}{s^2} - \frac{11}{s-3}\right] = 5\mathcal{L}^{-1}\left[\frac{1}{s^2}\right] - 11\mathcal{L}^{-1}\left[\frac{1}{s-3}\right]$$

This property will be very useful when solving differential equations later on.

8.3.4 Basic Relations

There are some basic relations that can be used when solving differential equation problems of which the most important ones are usually the Laplace transforms of the first and second derivative of functions. These are properties 6 and 7 in the Table 8.2 where a is an arbitrary constant. These relationships can be proved

Table 8.2 Basic Properties of Laplace Transforms

Operation	Time Domain	s-Domain
1. Time shifting	$f(t-a)$	$e^{-as}F(s)$
2. Time scaling	$f(at)$	$\dfrac{1}{a}F\left(\dfrac{s}{a}\right)$
3. Multiplying by an exponential in t-domain	$e^{at}f(t)$	$F(s-a)$
4. Multiplying by (t)	$tf(t)$	$-\dfrac{d}{ds}\left[F(s)\right]$
5. Dividing by (t)	$\dfrac{1}{t}f(t)$	$\displaystyle\int_s^\infty F(s)\,ds$
6. First derivative	$f'(t)$	$sF(s) - f(0)$
7. Second-order derivative	$f''(t)$	$s^2F - sf(0) - f'(0)$
8. Integration	$\displaystyle\int_0^t f(t)\,dt$	$\dfrac{1}{s}F(s)$

using the basic definition given by Equation 8.5. The following examples show how properties 1 and 6 of Table 8.2 are derived.

Example 8.8

Show that the $\mathcal{L}[f(t-a)] = e^{-as}F(s)$ (Property 1).

Using the formula given by Equation 8.5, $\mathcal{L}\left[f(t)\right] = \displaystyle\int_0^\infty f(t)e^{-st}\,dt$,

$$\mathcal{L}\left[f(t-a)\right] = \int_0^\infty f(t-a)e^{-st}\,dt \tag{8.7}$$

by replacing $f(t)$ with $f(t-a)$ and making a change of variable by letting $u = t - a$. This gives $du = dt$ and a is a constant and $t = u + a$. The limits of integration remain the same and making the above substitutions into Equation 8.7 gives

$$\mathcal{L}\left[f(t-a)\right] = \int_0^\infty f(u)e^{-s(u+a)}\,du = \int_0^\infty f(u)e^{-su}e^{-as}\,du$$

The e^{-as} term can be taken out in front of the integral since it does not depend on u.

$$\mathcal{L}\left[f(t-a)\right] = e^{-as}\int_0^\infty f(u)e^{-su}\,du$$

But $\displaystyle\int_0^\infty f(u)e^{-su}\,du = F(s)$, so the following result is obtained:

$$\mathcal{L}\left[f(t-a)\right] = e^{-as}F(s)$$

Example 8.9

Show that the $\mathcal{L}[f'(t)] = sF(s) - f(0)$ (Property 6).

Using the definition gives $\mathcal{L}[f'(t)] = \int\limits_0^\infty f'(t)e^{-st}\,dt$. Now this integral has

to done by using integration by parts. Letting

$$u = e^{-st} \quad \text{and} \quad \frac{dv}{dt} = f'(t)$$

gives

$$\frac{du}{dt} = -se^{-st} \quad \text{and} \quad v = f(t)$$

Therefore,

$$\mathcal{L}[f'(t)] = \left[e^{-st}f(t)\right]_0^\infty - \int\limits_0^\infty f(t)(-se^{-st})\,dt = \left[0 - f(0)\right] + s\int\limits_0^\infty f(t)e^{-st}\,dt$$

$$= -f(0) + s\,F(s)$$

giving the result that

$$\mathcal{L}[f'(t)] = s\,F(s) - f(0)$$

Note: This result together with the second derivative transform (property 7) are used extensively when solving differential equations.

8.4   Inverse Transforms

In the final part of the process of solving differential equations (see Figure 8.1), one needs to obtain the original function back in the time domain. This will require once again using the standard table of transforms and inverses. Usually, the expression that is required to be inverse Laplace transformed cannot be readily obtained from the table as it stands. However, with the use of *partial fraction decomposition* (see Chapter 6, Section 6.3.2) expressions can be obtained that can be inverse Laplace transformed easily.

Next, examples are given on how to find the inverse Laplace transform of some given functions.

Example 8.10

Find

$$\mathcal{L}^{-1}\left[\frac{2}{s^2(s^2+4)}\right].$$

This is not readily available as it stands in the standard tables. The fraction $\dfrac{2}{s^2(s^2+4)}$ has to be split into simpler fractions using partial fraction decomposition:

$$\frac{2}{s^2(s^2+4)}=\frac{A}{s}+\frac{B}{s^2}+\frac{Cs+D}{s^2+4} \qquad (8.8)$$

Multiplying throughout by the $s^2(s^2+4)$ term gives the following:

$$2 = A(s)(s^2+4)+B(s^2+4)+(Cs+D)s^2 \qquad (8.9)$$

Using easy values of s (i.e., $s = 0$) and then equating coefficients on both sides of Equation 8.8 gives

$$A=0, \quad B=\frac{1}{2}, \quad C=0, \quad \text{and} \quad D=-\frac{1}{2}$$

Therefore, Equation 8.8 can now be written as

$$\frac{2}{s^2(s^2+4)}=\frac{1}{2s^2}-\frac{1}{2(s^2+4)}$$

$$\mathcal{L}^{-1}\left[\frac{2}{s^2(s^2+4)}\right]=\mathcal{L}^{-1}\left[\frac{1}{2s^2}\right]-\mathcal{L}^{-1}\left[\frac{1}{2}\frac{1}{(s^2+4)}\right]$$

$$\mathcal{L}^{-1}\left[\frac{2}{s^2(s^2+4)}\right]=\frac{1}{2}\mathcal{L}^{-1}\left[\frac{1}{s^2}\right]-\frac{1}{2}\mathcal{L}^{-1}\left[\frac{1}{(s^2+4)}\right]$$

Finally, the inverse transforms are obtained from using the table of standard transforms:

$$\mathcal{L}^{-1}\left[\frac{2}{s^2(s^2+4)}\right]=\frac{1}{2}t-\frac{1}{4}\sin(2t)$$

Example 8.11

If $F(s) = \dfrac{s+4}{(s+3)(3s-2)}$, find $f(t)$.

Solution: First using the partial fractions method gives

$$\frac{s+4}{(s+3)(3s-2)} = \frac{A}{s+3} + \frac{B}{3s-2} \qquad (8.10)$$

Multiplying throughout by the $(s + 3)\ (3s - 2)$ term gives

$$s+4 = A(3s-2) + B(s+3)$$

Using easy values of s, that is, $s = -3$ and $s = \dfrac{2}{3}$, gives

$$A = -\frac{1}{11} \quad \text{and} \quad B = \frac{14}{11}$$

Therefore, Equation 8.10 becomes by writing the constants at the front as

$$\frac{s+4}{(s+3)(3s-2)} = \frac{-1}{11}\frac{1}{s+3} + \frac{14}{11}\frac{1}{3s-2}$$

and so

$$f(t) = \mathcal{L}^{-1}\left[F(s)\right]$$

$$f(t) = \mathcal{L}^{-1}\left[\frac{s+4}{(s+3)(3s-2)}\right] = \mathcal{L}^{-1}\left[\frac{-1}{11}\frac{1}{(s+3)}\right] + \mathcal{L}^{-1}\left[\frac{14}{11}\frac{1}{(3s-2)}\right]$$

$$f(t) = \frac{-1}{11}\mathcal{L}^{-1}\left[\frac{1}{(s+3)}\right] + \frac{14}{11}\mathcal{L}^{-1}\left[\frac{\frac{1}{3}}{\left(s-\frac{2}{3}\right)}\right]$$

Here in the term $(3s - 2)$ the 3 has been taken out of the bracket first and then using the standard table of transforms gives

$$f(t) = \frac{-1}{11}e^{-3t} + \frac{14}{33}e^{\frac{2}{3}t}$$

So far, it has been shown how to Laplace transform and inverse Laplace transform certain continuous time functions. However, in some engineering problems, the input to the system is *discontinuous* in time and hence consideration is now given on how to Laplace transform some of these kinds of input functions.

8.5 Discontinuous Functions

8.5.1 Heaviside Unit Step Function

The unit step function has the effect of switching on or switching off at some pre-described value of the time t. The function is shown in Figure 8.2. This function is the Heaviside unit step and is denoted by $f(t) = H(t - c)$, where

$$f(t) = 0 \quad :t < c$$
$$f(t) = 1 \quad :t \geq c$$

8.5.1.1 Calculating the Laplace Transform of H(t - c)

$$\mathcal{L}\big[H(t-c)\big] = \int_0^\infty H(t-c)e^{-st}\,dt$$

But $H(t-c)e^{-st} = 0 \qquad t < c$
$$= e^{-st} \qquad t \geq c$$

Therefore,

$$\mathcal{L}\big[\, H(t-c)\, \big] = \int_0^\infty H(t-c)e^{-st}\,dt = \int_c^\infty e^{-st}\,dt = \left[\frac{e^{-st}}{-s}\right]_c^\infty = 0 - \frac{e^{-sc}}{-s} = \frac{e^{-sc}}{s}$$

Generally,

$$\mathcal{L}\big[H(t-c)\big] = \frac{e^{-sc}}{s} \qquad\qquad (8.11)$$

This is the Laplace transform of unit step function operating at $t = c$.

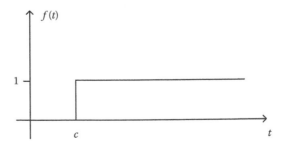

Figure 8.2 Heaviside unit step function operating at $t = c$.

8.5.1.2 Unit Step at Origin

If the unit step occurs at the origin, then putting $c = 0$ in $H(t - c)$ gives $f(t) = H(t)$ and this is shown in Figure 8.3.

Also, the Laplace transform of the unit step function at the origin becomes

$$\mathcal{L}\left[H(t)\right] = \frac{e^{-s0}}{s} = \frac{1}{s}$$

using $c = 0$ in Equation 8.11.

More generally, special kinds of discontinuous input functions can be described using the Heaviside unit step functions operating at different points. An example is the square pulse operating from $t = a$ to $t = b$ as given in Figure 8.4.

Here, we can represent $f(t)$ by subtracting two Heaviside unit step functions as

$$f(t) = H(t - a) - H(t - b)$$

Hence, the Laplace transform of the square pulse becomes

$$\mathcal{L}\left[f(t)\right] = L\left[H(t - a) - H(t - b)\right] = \frac{e^{-sa}}{s} - \frac{e^{-sb}}{s}$$

using Equation 8.11.

Next, another discontinuous function known as the delta (or impulse) function is considered.

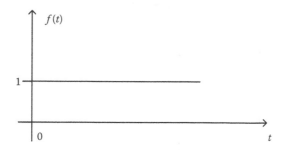

Figure 8.3 Heaviside unit step function operating at $t = 0$.

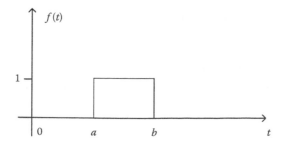

Figure 8.4 Square pulse operating between $t = a$ and $t = b$.

8.5.2   The Delta Function

Graphically the delta function is a rectangular pulse of zero width, infinite height, and can be represented by a single vertical line with an arrowhead as shown in Figure 8.5.

The definition of the delta function operating at $t = a$ is

$$\delta(t-a) = 0, \quad t \neq a$$
$$\delta(t-a) = \infty, \quad t = a$$

and the area of the pulse is 1.

The Laplace transform of $\delta(t-a)$ is then given by

$$\mathcal{L}\big[\delta(t-a)\big] = \int_0^\infty \delta(t-a)e^{-st}\,dt$$

$$= e^{-sa}\int_0^\infty \delta(t-a)\,dt$$

But $\displaystyle\int_0^\infty \delta(t-a)\,dt = 1$ (just the area of the delta function).

Therefore, the Laplace of the delta function operating at $t = a$ is

$$\mathcal{L}\big[\delta(t-a)\big] = e^{-sa} \tag{8.12}$$

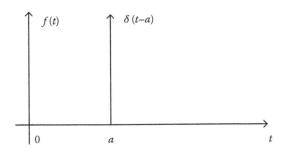

Figure 8.5 Delta function operating at $t = a$.

8.5.2.1   The Delta Function at the Origin

The delta function at the origin as shown in Figure 8.6 is defined as

$$\delta(t) = 0, \quad t \neq 0$$
$$\delta(t) = \infty, \quad t = 0$$

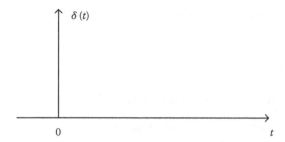

Figure 8.6 Delta function operating at $t = 0$.

It follows that the Laplace transform of the delta function at the origin with $a = 0$ is

$$\mathcal{L}\big[\delta(t)\big] = 1$$

using Equation 8.12.

Note: Use of these types of discontinuous input functions can arise in engineering problems. For example, in fire these can be used to determine what happens if a window breaks or a suppression system activates. Hence, knowing their Laplace Transforms enables solutions to problems as is shown later in the applications section (Section 8.8).

8.6   Shift Theorems

When working with differential equations with discontinuous input functions, the final part of inverse Laplace transforming sometimes requires the use of the first and second shift theorems and so these are stated next:

- First shift theorem (s-shifting)

$$f(t) \leftrightarrow F(s), \quad \text{then} \quad e^{at} f(t) \leftrightarrow F(s - a)$$

- Second shift theorem (t-shifting)

$$\text{If } F(s) = \mathcal{L}\big[f(t)\big], \quad \text{then} \quad e^{-cs} F(s) = \mathcal{L}\big[f(t - c)H(t - c)\big] \quad (8.13)$$

Again, these shift theorems can be derived using the definition formula of the Laplace transform. These are useful in some cases for finding inverse Laplace transforms, that is, taking the inverse Laplace transform of Equation 8.13 gives the following important result:

$$\mathcal{L}^{-1}\big[e^{-cs} F(s)\big] = f(t - c)H(t - c) \quad (8.14)$$

The following examples show how to make use of the second shift theorem to find inverses.

Example 8.12

Find a function $f(t)$ whose transform is $F(s) = \dfrac{e^{-4s}}{s^2}$.

Solution: The numerator e^{-4s} corresponds to e^{-cs} (i.e., $c = 4$) in Equation 8.14. Therefore, this indicates that there is a $H(t-4)$ term present and as

$$F(s) = \frac{1}{s^2}$$

implies that $f(t) = t$, which is the inverse Laplace transform of t.

$$\mathcal{L}^{-1}\left[\frac{e^{-4s}}{s^2}\right] = (t-4)H(t-4)$$

using Equation 8.14.

A systematic method for solving ordinary linear differential equations with a range of different inputs can now be outlined and followed as given next.

8.7   Method for Solving Linear Differential Equations

To solve linear differential equations by the Laplace Transform method, the following steps can be applied:

Step 1: Rewrite the differential equation in terms of the Laplace transform in the s-domain (i.e., Laplace transform every term).

Step 2: Insert the initial conditions.

Step 3: Rearrange the equation algebraically to give the Laplace transform of the solution.

Step 4: Finally, determine the inverse Laplace transform to obtain the solution (with the help of partial fractions and the table of standard transforms).

The following are key relationships that are used frequently in step 1:

$$\mathcal{L}\left[x(t)\right] = X(s)$$

$$\mathcal{L}\left[\frac{dx}{dt}\right] = sX(s) - x(0)$$

$$\mathcal{L}\left[\frac{d^2x}{dt^2}\right] = s^2X(s) - sx(0) - \dot{x}(0)$$

The next few examples show how to apply the preceding methodology to a range of different differential equations.

Example 8.13

Find the solution to the first-order differential equation

$$\frac{dx}{dt} + 3x = e^{-2t} \tag{8.15}$$

with given initial conditions that $x = 2$ when $t = 0$.

Solution:

Step 1: Laplace transforming Equation 8.15:

$$\mathcal{L}\left[\frac{dx}{dt}\right] + \mathcal{L}[3x] = \mathcal{L}\left[e^{-2t}\right]$$

$$sX(s) - x(0) + 3X(s) = \frac{1}{s+2}$$

Step 2: Insert initial conditions, that is, $x(0) = 2$:

$$sX(s) - 2 + 3X(s) = \frac{1}{s+2}$$

Step 3: Find $X(s)$ using algebraic manipulation

$$(s+3)X(s) = \frac{1}{s+2} + 2$$

$$X(s) = \frac{2s+5}{(s+2)(s+3)}$$

Step 4: Find the inverse Laplace transform

$$x(t) = \mathcal{L}^{-1}\left[X(s)\right] = \mathcal{L}^{-1}\left[\frac{2s+5}{(s+2)(s+3)}\right]$$

This is done by using partial fraction splitting of the fraction

$$\frac{2s+5}{(s+2)(s+3)}$$

$$x(t) = \mathcal{L}^{-1}\left[X(s)\right] = \mathcal{L}^{-1}\left[\frac{1}{(s+2)} + \frac{1}{(s+3)}\right] = \mathcal{L}^{-1}\left[\frac{1}{(s+2)}\right] + \mathcal{L}^{-1}\left[\frac{1}{(s+3)}\right]$$

Therefore, $x(t) = e^{-2t} + e^{-3t}$ using standard tables.

Example 8.14

Solve the following second-order differential equation:

$$\frac{d^2y}{dt^2} + 5\frac{dy}{dt} + 4y = 3\delta(t-2) \qquad (8.16)$$

with initial conditions $y(0) = 2$ and $y'(0) = -2$.

Solution:

Step 1: Laplace transforming Equation 8.16:

$$\mathcal{L}\left[\frac{d^2y}{dt^2}\right] + 5\mathcal{L}\left[\frac{dy}{dt}\right] + 4\mathcal{L}[y] = 3\,\mathcal{L}\big[\delta(t-2)\big]$$

$$s^2Y(s) - sy(0) - \dot{y}(0) + 5\big(sY(s) - y(0)\big) + 4Y(s) = 3e^{-2s}$$

Step 2: Insert initial conditions:

$$s^2Y(s) - 2s + 2 + 5(sY(s) - 2) + 4Y(s) = 3e^{-2s}$$

Step 3: Find $Y(s)$ using algebraic manipulation

$$(s^2 + 5s + 4)Y(s) - 2s - 8 = 3e^{-2s}$$

$$Y(s) = \frac{3e^{-2s} + 2s + 8}{(s^2 + 5s + 4)}$$

$$Y(s) = \frac{3e^{-2s} + 2(s+4)}{(s+1)(s+4)}$$

$$Y(s) = \frac{3e^{-2s}}{(s+1)(s+4)} + \frac{2}{(s+1)}$$

Step 4: Find the inverse Laplace transform. Splitting

$$\frac{3}{(s+1)(s+4)}$$

using partial fractions gives

$$Y(s) = \left[\frac{1}{(s+1)} - \frac{1}{(s+4)}\right]e^{-2s} + \frac{2}{(s+1)}$$

and hence

$$y(t) = \mathcal{L}^{-1}[Y(s)]$$

$$y(t) = \mathcal{L}^{-1}\left[\frac{1}{(s+1)} - \frac{1}{(s+4)}\right]e^{-2s} + 2\,\mathcal{L}^{-1}\left[\frac{2}{(s+1)}\right]$$

Therefore, using the second shift theorem and standard table gives the solution as

$$y(t) = \left[e^{-(t-2)} - e^{-4(t-2)}\right]H(t-2) + 2e^{-t}$$

Example 8.15

This is a more complicated example in which there are two-coupled second-order differential equations to solve.

Solve the following second-order simultaneous equations:

$$\frac{d^2x}{dt^2} + 2x - y = 0 \qquad\qquad (8.17)$$

$$\frac{d^2y}{dt^2} + 2y - x = 0 \qquad\qquad (8.18)$$

with initial conditions

$$x(0) = 4 \quad \text{and} \quad \dot{x}(0) = 0; \qquad y(0) = 2 \quad \text{and} \quad \dot{y}(0) = 0$$

Solution:

Step 1: Laplace transforming Equations 8.17 and 8.18:

$$\mathcal{L}\left[\frac{d^2x}{dt^2}\right] + 2\mathcal{L}[x] - \mathcal{L}[y] = L[0]$$

$$\mathcal{L}\left[\frac{d^2y}{dt^2}\right] + 2\mathcal{L}[y] - \mathcal{L}[x] = L[0]$$

$$s^2 X(s) - s\,x(0) - \dot{x}(0) + 2X(s) - Y(s) = 0$$

$$s^2 Y(s) - s\,y(0) - \dot{y}(0) + 2Y(s) - X(s) = 0$$

Step 2: Put in initial conditions

$$s^2 X(s) - 4s + 2\,X(s) - Y(s) = 0$$
$$s^2 Y(s) - 2s + 2\,Y(s) - X(s) = 0$$

Step 3: Simplifying algebraically

$$(s^2 + 2)X(s) - Y(s) = 4s \tag{8.19}$$

$$(s^2 + 2)Y(s) - X(s) = 2s \tag{8.20}$$

These are two simultaneous equations in $X(s)$ and $Y(s)$ and we can eliminate $Y(s)$ to obtain $X(s)$ first.

Multiplying Equation 8.19 by $(s^2 + 2)$ and then adding the two equations and after simplifying an expression for $X(s)$ can be found. Hence, solving the preceding two equations for $X(s)$ gives

$$X(s) = \frac{4s^3 + 10s}{(s^2 + 1)(s^2 + 3)}$$

Step 4: Inverse Laplace transform using partial fractions and tables:

$$X(s) = \frac{3s}{(s^2 + 1)} + \frac{s}{(s^2 + 3)}$$

Hence,

$$x(t) = \mathcal{L}^{-1}\left[\frac{3s}{(s^2 + 1)}\right] + \mathcal{L}^{-1}\left[\frac{s}{(s^2 + 3)}\right]$$

$$x(t) = 3\mathcal{L}^{-1}\left[\frac{s}{(s^2 + 1)}\right] + \mathcal{L}^{-1}\left[\frac{s}{\left(s^2 + \left(\sqrt{3}\right)^2\right)}\right]$$

$$x(t) = 3\cos(t) + \cos\left(\sqrt{3}t\right) \tag{8.21}$$

Now $y(t)$ can be found using the original Equation 8.17. Using Equation 8.17 and rearranging gives

$$y(t) = \frac{d^2x}{dt^2} + 2x \tag{8.22}$$

Differentiating $x(t)$ (i.e., Equation 8.21) once gives

$$\frac{dx}{dt} = -3\sin t - \sqrt{3}\sin\sqrt{3}t$$

Differentiating this again gives

$$\frac{d^2x}{dt^2} = -3\cos t - 3\cos\sqrt{3}t$$

Then using Equation 8.22 gives the solution for $y(t)$ as

$$y(t) = 3\cos(t) - \cos\left(\sqrt{3}t\right)$$

In many applications, there is a relationship between two variables and the interdependence of the variables produces coupled differential equations. In fire engineering, the variables are usually the temperature and the fire radius, which then generate a coupled differential equation when modeling the phenomenon. Modeling predator–prey relationships of simple ecosystems often produces coupled first-order differential equations. However, in this case the equations are nonlinear in nature and generally more difficult to solve.

Having seen how the Laplace transform method can be applied to solving general differential equations, in the next section a range of different real-world applications are given where the Laplace transform method can be used to solve the problem.

8.8 Applications

Example 8.16: Temperature Variation of the Hot Gas and Smoke Layer

A simple two-zone model of a room splits the room into an upper smoke layer and a lower layer with the fire as shown in Figure 8.7.

A simple equation for a two-zone model of a fire in a room can be derived from using the heat balance equation:

Heat in the room = Heat generated from the fire – Heat losses in the room

which gives the following equation:

$$mc_p \frac{dT}{dt} = \dot{Q} - Ah\Delta T \tag{8.23}$$

assuming that the initial temperature in the room is T_0 (i.e., $T(0) = T_0$) and if radiation is not a dominant phenomenon (i.e., nonflashover fires).

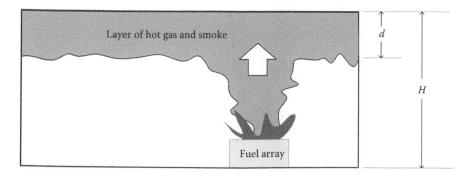

Figure 8.7 Two-zone model of a fire in an enclosed room.

The differential Equation 8.23 can be simplified by dividing by mc_p and letting $\Delta T = T - T_0$ as follows:

$$\frac{dT}{dt} = \frac{\dot{Q}}{mc_p} - \frac{Ah}{mc_p}(T - T_0) \tag{8.24}$$

Simplification can be done further by making the following substitutions for the constants as

$$a = \frac{Ah}{mc_p} \quad \text{and} \quad b = \frac{\dot{Q}}{mc_p} + aT_0$$

into Equation 8.24. This gives the following simple first-order differential equation:

$$\frac{dT}{dt} = b - aT \tag{8.25}$$

with initial condition that $T(0) = T_0$ to solve.

This differential Equation 8.25 can be solved to find how the temperature T of the hot smoke varies as a function of time t using the Laplace transform method.

Solution: Starting with Equation 8.25 and first rearranging gives

$$\frac{dT}{dt} + aT = b \quad \text{with} \quad T(0) = T_0 \tag{8.26}$$

Step 1: Laplace transforming Equation 8.26:

$$\mathcal{L}\left[\frac{dT}{dt}\right] + a\mathcal{L}[T] = \mathcal{L}[b]$$

$$sT(s) - T(0) + a\,T(s) = \frac{b}{s}$$

Step 2: Substitute in the initial condition:

$$sT(s) - T_0 + aT(s) = \frac{b}{s}$$

Step 3: Rearranging algebraically to find $T(s)$:

$$sT(s) + aT(s) = \frac{b}{s} + T_0$$

$$(s+a)T(s) = \frac{b + sT_0}{s}$$

$$T(s) = \frac{sT_0 + b}{s(s+a)}$$

Step 4: Using partial fraction decomposition and inverse Laplace transforms gives

$$\frac{sT_0 + b}{s(s+a)} = \frac{A}{s} + \frac{B}{s+a} \tag{8.27}$$

Multiplying throughout by $s(s + a)$ gives

$$sT_0 + b = A(s+a) + Bs$$

Using $s = 0$ gives $A = \dfrac{b}{a}$ and using $s = -a$ gives $B = T_0 - \dfrac{b}{a}$. This gives
Equation 8.27 as

$$\frac{sT_0 + b}{s(s+a)} = \frac{b}{a}\frac{1}{s} + \left(T_0 - \frac{b}{a}\right)\frac{1}{s+a}$$

So,

$$T(s) = \frac{b}{a}\frac{1}{s} + \left(T_0 - \frac{b}{a}\right)\frac{1}{s+a} \tag{8.28}$$

Applying the inverse Laplace transform to Equation 8.28 to give $T(t)$
using the standard tables as

$$T(t) = \mathcal{L}^{-1}\left[T(s)\right]$$

$$T(t) = \mathcal{L}^{-1}\left[\frac{b}{a}\frac{1}{s} + \left(T_0 - \frac{b}{a}\right)\frac{1}{s+a}\right] = \frac{b}{a}\mathcal{L}^{-1}\left[\frac{1}{s}\right] + \left(T_0 - \frac{b}{a}\right)\mathcal{L}^{-1}\left[\frac{1}{s+a}\right]$$

$$T(t) = \frac{b}{a}\left(T_0 - \frac{b}{a}\right)e^{-at}$$

Substituting back for the constants a and b and collecting terms gives
$T(t)$:

$$T(t) = T_0 e^{-\frac{Ah}{mc_p}t} + \left(T_0 + \frac{\dot{Q}}{Ah}\right)\left(1 - e^{-\frac{Ah}{mc_p}t}\right) \tag{8.29}$$

In Equation 8.29, as $t \to \infty$ the exponential terms $\to 0$, giving the long-
term temperature as $T(\infty) \to T_0 + \dfrac{\dot{Q}}{Ah}$.

The solution can be shown graphically as in Figure 8.8.

It can be seen that the temperature in the room increases to a steady-
state value as long as there is still fuel burning. The fire will grow toward

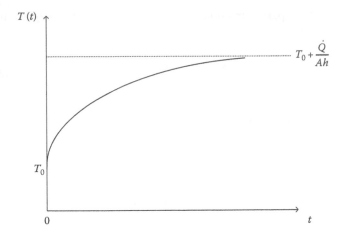

Figure 8.8 Graph showing temperature variation against time.

a "quasi-steady" temperature. It will keep settling closer and closer to this value until the fuel source eventually runs out. When $\dot{Q} = 0$, then the temperature will start to decay again. So, Equation 8.24 can now be solved with $\dot{Q} = 0$, and the approximate initial conditions produce an *exponential decay* for the temperature distribution once the fuel has run out.

Example 8.17: Basic Fire Growth Model Using Unit Step Functions

As an example, in fire combustion a basic *crude* model of fire growth is that it can be represented as a pulse wave similar to that given in Figure 8.4 with the heat release rate $\dot{Q}(t)$ against time t shown in Figure 8.9.

Here in the first 2 minutes it is assumed that the fire is just getting started and then there is a constant fire for around 20 minutes before it dies out. The heat release above could be represented in terms of the unit step functions with the units here being MW as follows:

$$\dot{Q}(t) = H(t-2) - H(t-22)$$

The next example shows how the unit step function can be used to represent more general functions and their applications.

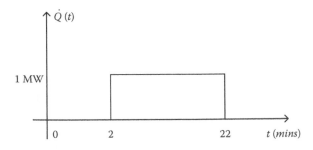

Figure 8.9 Heat release rate against time.

Example 8.18: Piecewise Function Representation

In some practical situations, a function $f(t)$ may wary differently over time as follows:

$$f(t) = \begin{cases} f_1(t) & 0 \le t < t_1 \\ \\ f_2(t) & t_1 \le t \le t_2 \end{cases}$$

Graphically, the function may be represented as shown in Figure 8.10.

However, again using the definition of the unit step function, this function $f(t)$ can now be represented as a single function as follows:

$$f(t) = \left[1 - H(t - t_1)\right]f_1(t) + \left[H(t - t_1) - H(t - t_2)\right]f_2(t) \qquad (8.30)$$

Since when $0 < t < t_1$, then $H(t - t_1) = 0$ and $H(t - t_2) = 0$. Then $f(t) = f_1(t)$.

When $t_1 \le t < t_2$, then $H(t - t_1) = 1$, $H(t - t_2) = 0$, and $f(t) = f_2(t)$.

And when $t \ge t_2$, then now $H(t - t_1) = 1$ and $H(t - t_2) = 1$, so $f(t) = 0$ again as required.

This representation of $f(t)$ is true for all times.

Figure 8.10 can now be used to model a more realistic development of fire growth then that given in Figure 8.9 in which the first function $f_1(t)$ is a t-squared function and the second function $f_2(t)$ is a constant K. Now, the value of t_1 would vary for the different types of fires from 10 minutes for a slow fire, 5 minutes for a medium, 2.5 minutes for a fast, and 1.25 minutes for an ultrafast fire.

If $f(t)$ is now the heat release rate $\dot{Q}(t)$, then this can be represented by a single function as follows using Equation 8.30:

$$\dot{Q}(t) = \left[1 - H(t - t_1)\right]t^2 + \left[H(t - t_1) - H(t - t_2)\right]K$$

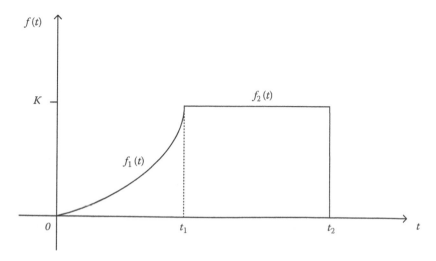

Figure 8.10 A piecewise function over time.

In other engineering fields, the use of the unit step function can offer a method to derive solutions for problems in which the boundary conditions depend upon time. An example of this is found in oil and gas exploration where there is a pressure buildup following constant pressure production. The boundary conditions for the oil well are now different for the constant pressure and for the pressure buildup regions. Representing the boundary conditions as a single function using the unit step functions as in Equation 8.30 allows solutions to be found subsequently using the Laplace Transform method. This scenario often arises in drill stem testing of gas wells.

Example 8.19: Transient Analysis to Determine the Current $i(t)$ in a RCL Circuit

In Figure 8.11 is an electrical circuit with a voltage source $V(t)$, a resistor with resistance R, an inductor with inductance L, and a capacitor with capacitance C all connected in series.

Using Kirchhoff's law for the voltage around the circuit, it is known that the voltage source $V(t)$ is equal to the voltage across the resistor, inductor, and the capacitor. In terms of the current $i(t)$ and charge $q(t)$ this yields the following first-order differential equation:

$$L\frac{di}{dt} + Ri + \frac{q}{C} = V \tag{8.31}$$

Since the current in the circuit is given by

$$i(t) = \frac{dq}{dt} \tag{8.32}$$

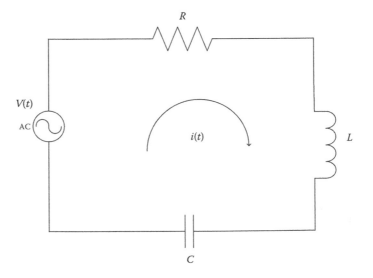

Figure 8.11 *RCL* electrical circuit.

this implies that $\dfrac{di}{dt} = \dfrac{d^2q}{dt^2}$. So Equation 8.31 can be written in terms of the charge $q(t)$ by replacing for $\dfrac{di}{dt}$ and for $i(t)$ as follows:

$$L\frac{d^2q}{dt^2} + R\frac{dq}{dt} + \frac{q}{C} = V \qquad (8.33)$$

This differential equation has initial conditions for the current and the charge in the circuit, which can be written as $\dot{q}(0)$ and $q(0)$, respectively.

This electrical circuit problem can now be solved using the Laplace transform method to first find the charge $q(t)$ and then subsequently the current in the circuit $i(t)$ using Equation 8.32.

Solution:

Step 1: Laplace transforming Equation (8.33):

$$\mathcal{L}\left[L\frac{d^2q}{dt^2} \right] + R\,\mathcal{L}\left[\frac{dq}{dt} \right] + \mathcal{L}\left[\frac{q}{c} \right] = \mathcal{L}[V]$$

$$L\left[s^2 Q(s) - s\,q(0) - \dot{q}(0) \right] + R\left[s\,Q(s) - q(0) \right] + \frac{1}{C}Q(s) = V(s)$$

Step 2: The initial conditions are just $q(0)$ and $\dot{q}(0)$ here and so can be left as such.

Step 3: Rearrange the equation algebraically to find $Q(s)$.

$$\left(L\,s^2 + Rs + \frac{1}{C} \right)Q(s) = V(s) + Ls\,q(0) + L\,\dot{q}(0) + R\,q(0)$$

$$Q(s) = \frac{V(s) + Ls\,q(0) + L\,\dot{q}(0) + R\,q(0)}{L\,s^2 + Rs + \dfrac{1}{C}} \qquad (8.34)$$

Step 4: Inverse Laplace transform gives

$$q(t) = \mathcal{L}^{-1}\left[Q(s) \right] = \mathcal{L}^{-1}\left[\frac{V(s) + Ls\,q(0) + L\,\dot{q}(0) + R\,q(0)}{L\,s^2 + Rs + \dfrac{1}{C}} \right] \qquad (8.35)$$

The initial conditions $q(0)$, $\dot{q}(0)$ are known and so is the voltage source $V(t)$. Hence $V(s)$ is known. Also, L, R, and C are all constants, which are all known.

So, from Equation 8.35, the charge $q(t)$ in the circuit can be found. The current $i(t)$ can then be found from Equation 8.32 using $i(t) = \dfrac{dq}{dt}$, that is, by differentiating the expression for the charge $q(t)$.

Note: An important part of fire studies is fire investigation. After the fire has been put out by firefighters, fire investigators have to determine what was the original cause of the fire. Most fire investigation studies carried out are insurance related, where the cause of the fire can often have an electrical cause as the suspicion. For example, with a voltage source V = 230 V and a device that has been running for say 2 hours, the temperatures reached can be calculated approximately using the appropriate energy equations, that is, VIt = mcΔT. So, understanding well the basic concepts of electrical flow are essential to the fire investigators' knowledge base.

Example 8.20: Application in Control Systems

In control systems, the output response of a system is designed to relate in a particular way to the input of the system. It works by comparing the input response of the system to the output response using appropriate feedback control so as to reduce the difference between the input and the output to zero.

For a general component of a control system, the transfer function $H(s)$ relates the Laplace transform $Y(s)$ of the output $y(t)$ to the Laplace transform $R(s)$ of the input $r(t)$ as seen in Figure 8.12.

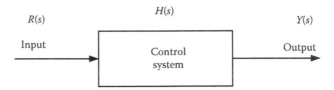

Figure 8.12 Control system representation $Y(s) = H(s) R(s)$.

A general symbolic representation of a control feedback system is shown in Figure 8.13.

Using some basic block diagram representation, this gives $E(s) = R(s) - Z(s)$.

$$Y(s) = G(s) E(s) = G(s)\left[R(s) - Z(s)\right] \tag{8.36}$$

So, $Y(s) = G(s) E(s) = G(s) [R(s) - Z(s)]$.

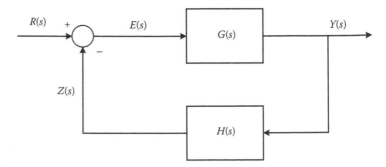

Figure 8.13 Control feedback system representation.

But $Z(s) = H(s) Y(s)$. Substituting this in above for $Z(s)$ gives

$$Y(s) = G(s)\left[R(s) - H(s)Y(s)\right]$$

Rearranging to find $Y(s)$ gives

$$Y(s) = G(s)R(s) - G(s)H(s)Y(s)$$
$$Y(s) + G(s)H(s)Y(s) = G(s)R(s)$$
$$Y(s)\left[1 + G(s)H(s)\right] = G(s)R(s)$$

$$Y(s) = \frac{G(s)}{\left[1 + G(s)H(s)\right]}R(s) \qquad (8.37)$$

This can be written as

$$Y(s) = T(s)R(s)$$

where

$$T(s) = \frac{G(s)}{\left[1 + G(s)H(s)\right]}$$

Thus, the transfer function for the control system is $T(s)$ in which the feedback transfer function $H(s)$ can be manipulated for design purposes as required.

The solution for Equation 8.37 can be found by the Laplace transforms and inverse Laplace transforms method once $G(s)$, $H(s)$, and $R(s)$ are known.

$$y(t) = \mathcal{L}^{-1}\left[Y(s)\right] = \mathcal{L}^{-1}\left[\frac{G(s)}{\left[1 + G(s)H(s)\right]}R(s)\right]$$

This approach is of great importance in many real-world applications in a variety of different engineering disciplines, such as automated aircraft control. The main task here is ensuring that the feedback design produces a stable solution, that is, the aircraft remains in the desired position during flight.

Note: Although not studied extensively through this approach, flashover room fires can also be understood as thermal feedback problems.

Problems

8.1 Using the formula definition, find the Laplace transform $F(s)$ of $f(t) = t^2$. (Hint: Use integration by parts twice.)

8.2 Solve the following differential equations by using the Laplace transformation:

a. $y'(t) - 7\,y(t) = e^{-2t}$;　　　　　$y(0) = 1$

b. $T'(t) - 6\,T(t) = 3$;　　　　　$T(0) = 1$

c. $y''(t) - 9\,y'(t) + 8\,y(t) = 8$;　　$y(0) = 2,\ \dot{y}(0) = 1$

d. $x''(t) - 18x'(t) + 6 = 0$;　　　$x(0) = 2,\ \dot{x}(0) = 1$

8.3 Consider the following electrical RL circuit as shown in Figure 8.14, where $V(t)$ is the voltage source and a resistor with resistance R combined with an inductor with inductance L are connected in series. Kirchhoff's voltage law gives the differential equation for the current in the circuit as follows:

$$L\frac{di}{dt} + Ri = V(t)$$

Find an expression for the current $i(t)$ in the circuit given that $i(0) = 0$ for the following cases:

a. If $V(t) = V_0$ (i.e., constant).

b. If $V(t) = V_0 \sin wt$.

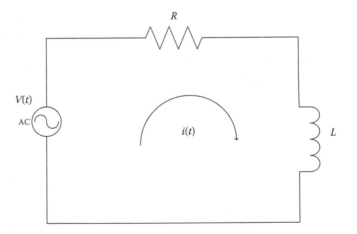

Figure 8.14 RL circuit.

8.4 Solve the following coupled first-order differential equations:

$$x'(t) - k\,y(t) = C$$

$$y'(t) - k\,x(t) = 0$$

with initial conditions $x(0) = y(0) = 0$, leaving your answers in terms of the constants k and C.

9 Fourier Series and Fourier Transforms

The Fourier series is named after the French mathematician Jean-Baptiste Joseph Fourier. A Fourier series is a mathematical way of representing a wavelike function (or signal) as a sum of simple sine and cosine waves. It decomposes a periodic function or periodic signal into a sum of an infinite set of sinusoidal functions. When these sines and cosines are expressed as complex exponentials this gives the Fourier series in complex form. As seen in Section 9.4 the Fourier transform is the generalization to nonperiodic functions. Since the Fourier series deals with periodic phenomena, it is important to first understand what is meant by a function being periodic in nature.

9.1 Periodic Functions

A periodic function (or signal) $f(x)$ is said to have a period T or be periodic with period T if for all values of x

$$f(x+T) = f(x) \qquad (9.1)$$

where T is a positive constant. The function then just repeats itself with period T over the whole interval $-\infty < x < \infty$.

Note: The number of oscillation per second, that is, frequency (Hz) $f = \dfrac{1}{T}$, and angular frequency (radians per second) is $w = \dfrac{2\pi}{T} = 2\pi f$.

The function $\sin x$ is an example of a periodic function with period 2π. The function repeats itself again after an interval of 2π (Figure 9.1).

Similar trigonometric functions that are periodic are the cosine function $\cos x$ with period 2π, and the tangent function $\tan x$ with period π.

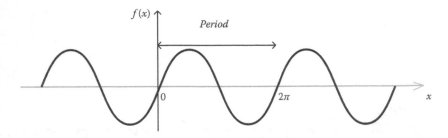

Figure 9.1 Sine wave function or signal.

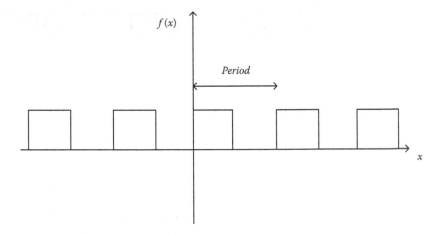

Figure 9.2 Square wave function.

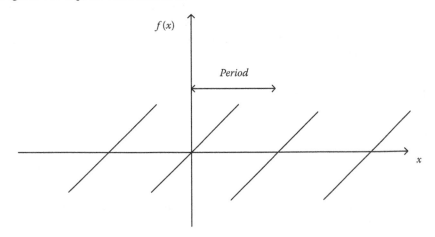

Figure 9.3 Sawtooth wave function.

Other examples of periodic functions are the square wave and the sawtooth function shown in Figures 9.2 and 9.3, respectively.

9.2   Fourier Series

9.2.1   Periodic Functions of Period T

Repeating functions can have different values for the period. For example, the sine function has period 2π A more general treatment of the Fourier series is

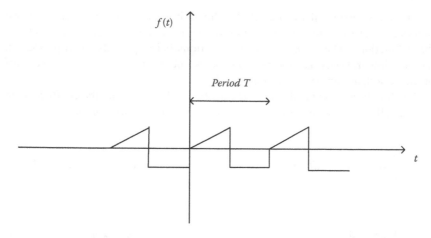

Figure 9.4 General periodic waveform with period T.

given where the period can be of any value, say, T, and considering the interval as being time t, a general periodic function $f(t)$ can be represent as in Figure 9.4. Here, $f(t + T) = f(t)$. Generally, $w = \dfrac{2\pi}{T}$, where w is the angular frequency (radians per second) and T is the period (seconds).

Using the idea that a periodic function can be represented as an infinite sum of sinusoidal functions, it can be shown that the Fourier series to represent $f(t)$ can be written as follows:

$$f(t) = \frac{a_0}{2} + \sum_{n=1}^{\infty} a_n \cos nwt + b_n \sin nwt \tag{9.2}$$

Notes:

- *When the period $T = 2\pi$, that is, $w = 1$, the terms in Equation 9.2 are made up of $\cos nt$ and $\sin nt$, which are periodic on the interval 2π for any integer n.*
- *The coefficients a_n and b_n measure the strength of the contribution from each "harmonic" in the series.*

The task is to see if $f(t)$ can be written in the form given by Equation 9.2, then what the coefficients a_0, a_n, and b_n have to be.

Before these can be found, it is important to have some understanding of orthogonality of functions as well as other general properties. These are covered in the next section.

9.2.2 General Properties and Orthogonal Functions

It is the case that the functions $\cos nt$ and $\sin nt$ have the following property that if they are integrated over a period then the result is zero, that is,

$$\int_{-\pi}^{\pi} \sin nt \ dt = \int_{-\pi}^{\pi} \cos nt \ dt = 0, \qquad \text{for all integers } n \tag{9.3}$$

These last two results can easily be shown by integrating the preceding two functions over the limits of integration but can also be seen from the graphs of these functions. The $\cos nt$ and $\sin nt$ functions being periodic with period 2π, then the integral over half the period cancels out the integral over the other half of the period, as can be seen in Figure 9.1.

The functions $\cos nt$ and $\sin nt$ can now be used to help find the coefficients in Equation 9.2 because they satisfy the following *orthogonality* properties:

$$\int_{-\pi}^{\pi} \sin mt \cos nt \ dt = 0, \qquad\qquad \text{for all } m \text{ and } n \qquad (9.4)$$

$$\int_{-\pi}^{\pi} \sin mt \sin nt \ dt = 0, \qquad\qquad \text{for } m \neq n$$
$$= \pi \quad \text{(half the period)}, \qquad \text{for } (m = n) > 0 \qquad (9.5)$$

$$\int_{-\pi}^{\pi} \cos mt \cos nt \ dt = 0, \qquad\qquad \text{for } m \neq n$$
$$= 2\pi \quad \text{(period)}, \qquad\quad \text{for } m = n = 0 \qquad (9.6)$$
$$= \pi \quad \text{(half the period)}, \qquad \text{for } (m = n) > 0$$

The orthogonality properties given in Equations 9.4, 9.5, and 9.6 can all be proved by either considering the graphs of the product functions or by expressing the product functions in terms of the sums of individual sine and cosine functions and integrating out (see trigonometric functions and integration sections). These results will all be very useful when calculating the Fourier coefficients in the next section.

Another important result for the Fourier series in complex form is that in the exponential notation the orthogonality conditions where m and n are integers become

$$\int_{-\pi}^{\pi} e^{jnt} e^{-jmt} \ dt = 0; \qquad\qquad m \neq n$$
$$= 2\pi \quad \text{(period)}; \qquad m = n \qquad (9.7)$$

Again, the proof of this is evident from direct integration and putting in the limits.

9.2.3   Fourier Coefficients

Starting with Equation 9.2 for the Fourier series:

$$f(t) = \frac{a_0}{2} + \sum_{n=1}^{\infty} a_n \cos nwt + b_n \sin nwt$$

The coefficients a_0, a_n, and b_n can be found by multiplying $f(t)$ by 1, $\cos mwt$, and $\sin mwt$, and integrating over a period, respectively, to give the following results:

$$a_0 = \frac{2}{T} \int_T f(t) \, dt \qquad (9.8)$$

$$a_n = \frac{2}{T} \int_T f(t) \cos nwt \, dt \qquad (9.9)$$

$$b_n = \frac{2}{T} \int_T f(t) \sin nwt \, dt \qquad (9.10)$$

Proofs:

For the coefficient a_0, starting with $f(t)$

$$f(t) = \frac{a_0}{2} + \sum_{n=1}^{\infty} a_n \cos nwt + b_n \sin nwt$$

integrating over the period T gives

$$\int_T f(t) \, dt = \int_T \frac{a_0}{2} \, dt + \sum_{n=1}^{\infty} \left[a_n \int_T \cos nwt \, dt + b_n \int_T \sin nwt \, dt \right]$$

Using Equation 9.3 gives this as

$$\int_T \frac{a_0}{2} \, dt = \int_T f(t) \, dt \qquad (9.11)$$

since

$$\int_T \cos nwt \, dt = \int_T \sin nwt \, dt = 0$$

Integrating the left-hand side of Equation 9.11 gives

$$\frac{a_0 T}{2} = \int_T f(t) \, dt$$

and so a_0 is

$$a_0 = \frac{2}{T} \int_T f(t) \, dt$$

as required.

Similarly, proofs for a_n and b_n can be shown using the orthogonality properties.

Note: For a general period T, it is better to first sketch the function f(t), then choose the appropriate periodic interval, that is, from 0 to T or $-\dfrac{T}{2}$ *to* $\dfrac{T}{2}$.

As an example of finding the trigonometric Fourier series of a periodic function, the next example will show how to calculate the coefficients and what the series looks like that represents the function.

Example 9.1

Determine the Fourier series for the periodic function defined by:

$$f(t) = \begin{cases} 2(1+t) & -1 < t < 0 \\ 0 & 0 < t < 1 \end{cases} \tag{9.12}$$

where $f(t) = f(t+2)$, that is, the period $T = 2$.

Solution: To see what the function given by Equation 9.12 looks like, it is first sketched as shown in Figure 9.5.

Starting with the general Fourier series for $f(t)$,

$$f(t) = \frac{a_0}{2} + \sum_{n=1}^{\infty} a_n \cos nwt + b_n \sin nwt \tag{9.13}$$

Since the period of $f(t)$ is equal to $T = 2$, this implies that $w = \dfrac{2\pi}{T} = \pi$.

Calculating the coefficient for a_0 using Equation 9.8 gives

$$a_0 = \frac{2}{T} \int_T f(t)\,dt = \frac{2}{2} \int_{-1}^{1} f(t)\,dt = \int_{-1}^{1} f(t)\,dt$$

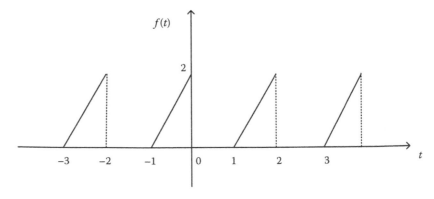

Figure 9.5 Graph of the function $f(t)$.

Now this integral has to be split into two regions since the function $f(t)$ is different in the two regions $-1 < t < 0$ and $0 < t < 1$, as can be seen from Figure 9.5.

$$a_0 = \int_{-1}^{1} f(t)\,dt = \int_{-1}^{0} f(t)\,dt + \int_{0}^{1} f(t)\,dt \tag{9.14}$$

Putting in the function values for the function in the different regions into Equation 9.14 gives

$$a_0 = \int_{-1}^{0} 2(1+t)\,dt + \int_{0}^{1} 0\,dt$$

$$a_0 = \int_{-1}^{0} 2(1+t)\,dt = 1$$

$a_0 = 1$ in Equation 9.13 for $f(t)$.

Calculating the coefficient for a_n using Equation 9.9 gives

$$a_n = \frac{2}{T} \int_{T} f(t)\cos nwt\,dt$$

Using $w = \pi$ gives

$$a_n = \frac{2}{2} \int_{-1}^{1} f(t)\cos n\pi t\,dt = \int_{-1}^{0} f(t)\cos n\pi t\,dt$$

Since $f(t)$ is zero otherwise

$$a_n = \int_{-1}^{0} 2(1+t)\cos n\pi t\,dt \tag{9.15}$$

Now this integral is a product of two functions and has to be performed by using integration by parts, that is, let $u = 2(1 + t)$ and then $\dfrac{dv}{dt} = \cos n\pi t$ gives

$$a_n = \frac{2}{n^2\pi^2}[1 - \cos n\pi] \tag{9.16}$$

where $n = 1, 2, 3, \ldots$.

This coefficient a_n has different values according to whether n is odd or even as follows:

$$\text{If } n = \text{odd } (1,3,5,\ldots), \quad a_n = \frac{4}{n^2\pi^2} \quad \text{because } \cos n\pi = -1 \qquad (9.17)$$

$$\text{If } n = \text{even } (2,4,6,\ldots), \quad a_n = 0 \quad \text{because } \cos n\pi = 1 \qquad (9.18)$$

These are the values of a_n that will be used in Equation 9.13 for $f(t)$.

Similarly, calculating for the coefficient for b_n using Equation 9.10 gives

$$b_n = \frac{2}{T} \int_T f(t) \sin nwt \, dt$$

with $w = \pi$, gives

$$b_n = \frac{2}{2} \int_{-1}^{1} f(t) \sin n\pi t \, dt = \int_{-1}^{0} f(t) \sin n\pi t \, dt$$

Since $f(t)$ is zero otherwise

$$b_n = \int_{-1}^{0} 2(1+t) \sin n\pi t \, dt \qquad (9.19)$$

Now this integral is again a product of two functions and has to be done by using integration by parts, that is, let $u = 2(1 + t)$ and then $\dfrac{dv}{dt} = \sin n\pi t$ gives

$$b_n = -\frac{2}{n\pi} \qquad (9.20)$$

where $n = 1, 2, 3, \ldots$.

Putting together all the coefficients found gives

$$a_0 = 1$$
$$a_n = \frac{4}{n^2\pi^2}, \qquad n = \text{odd}$$
$$a_n = 0, \qquad n = \text{even} \qquad (9.21)$$
$$b_n = -\frac{2}{n\pi}$$

Having all the coefficients, the Fourier series for $f(t)$ in Equation 9.2 can be written as

$$f(t) = \frac{1}{2} + \frac{4}{\pi^2}\left(\cos \pi t + \frac{\cos 3\pi t}{9} + \frac{\cos 5\pi t}{25} + \ldots\right)$$
$$- \frac{2}{\pi}\left(\sin \pi t + \frac{\sin 2\pi t}{2} + \frac{\sin 3\pi t}{3} + \ldots\right) \quad (9.22)$$

and in a more compact form $f(t)$ can be expressed as

$$f(t) = \frac{1}{2} + \sum_{n=odd}^{\infty}\left(\frac{4}{n^2\pi^2}\right)\cos n\pi t + \sum_{n=1}^{\infty}\left(-\frac{2}{n\pi}\right)\sin n\pi t \quad (9.23)$$

The above sum continues to an infinite number of terms. It can be seen how it converges to the original function by plotting a truncated sum of a finite number of terms. If the sum containing n-trigonometric terms is defined as $f_n(t)$, then

$$f_0(t) = \frac{1}{2}$$ (which is just the average value of the function over the period)

$$f_1(t) = \frac{1}{2} + \frac{4}{\pi^2}\cos \pi t - \frac{2}{\pi}\sin \pi t$$

$$f_2(t) = \frac{1}{2} + \frac{4}{\pi^2}\cos \pi t - \frac{2}{\pi}\sin \pi t - \frac{2}{2\pi}\sin 2\pi t, \text{ etc.}$$

The graphs of these functions can now be plotted to see how the series converges to the original function, as shown in Figure 9.6.

As more and more terms in the series are taken, the resulting function approximates the original signal more closely. This can be seen in Figure 9.6 with $n = 40$. The series is starting to approximate the original function given by Equation 9.12 and Figure 9.5 reasonably well.

Note: The oscillations seen in Figure 9.6b and c do become smaller and smaller as n gets larger and larger but do not disappear altogether since a discontinuous function is being represented by smooth sinusoidal functions. This is known as Gibbs phenomenon after J.W. Gibbs.

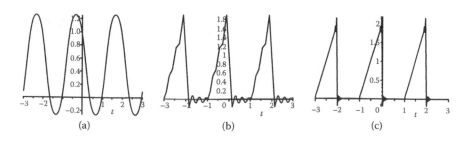

(a)   (b)   (c)

Figure 9.6 Graphs of $f_n(t)$ for values of n: (a) $n = 1$, (b) $n = 6$, and (c) $n = 40$.

9.3   Complex Form of the Fourier Series

Another way of representing the Fourier series is in terms of the complex exponentials and this turns out to have many useful applications in engineering.

It can be shown that the Fourier series given by Equations 9.2

$$f(t) = \frac{a_0}{2} + \sum_{n=1}^{\infty} a_n \cos nwt + b_n \sin nwt \tag{9.24}$$

can be written in a more compact form known as the *complex form*, which simplifies the calculations and is more useful especially when the Fourier transform is considered later.

Starting with Euler's formula:

$$e^{j\theta} = \cos\theta + j\sin\theta \tag{9.25}$$

then

$$e^{-j\theta} = \cos\theta - j\sin\theta \tag{9.26}$$

By adding and subtracting Equations 9.25 and 9.26 gives the following formulae for the cosine and sine function terms of exponentials:

$$\cos\theta = \frac{e^{j\theta} + e^{-j\theta}}{2} \tag{9.27}$$

$$\sin\theta = \frac{e^{j\theta} - e^{-j\theta}}{2j} \tag{9.28}$$

Letting $\theta = nwt$, gives the following expressions:

$$\cos nwt = \frac{e^{jnwt} + e^{-jnwt}}{2} \tag{9.29}$$

$$\sin nwt = \frac{e^{jnwt} - e^{-jnwt}}{2j} \tag{9.30}$$

In Equation 9.24 for the trigonometric Fourier series, the term inside the $\sum$, which is $a_n \cos nwt + b_n \sin nwt$, now becomes

$$\frac{1}{2}a_n\left(e^{jnwt} + e^{-jnwt}\right) + \frac{1}{2j}b_n\left(e^{jnwt} - e^{-jnwt}\right)$$

$$\frac{1}{2}\left(a_n - jb_n\right)e^{jnwt} + \frac{1}{2}\left(a_n + jb_n\right)e^{-jnwt} \tag{9.31}$$

Letting

$$c_n = \frac{1}{2}\left(a_n - jb_n\right) \text{ and } k_n = \frac{1}{2}\left(a_n + jb_n\right) \tag{9.32}$$

then

$$a_n \cos nwt + b_n \sin nwt = c_n e^{jnwt} + k_n e^{-jnwt} \tag{9.33}$$

Using Equation 9.33 into Equation 9.24 gives

$$f(t) = c_0 + \sum_{n=1}^{\infty} c_n e^{jnwt} + k_n e^{-jnwt} \tag{9.34}$$

where now c_0 is given as $c_0 = \dfrac{a_0}{2}$.

Also, using the definitions of c_n and k_n given in Equation 9.32, it can be shown further that $k_n = c_{-n}$, where c_n is given by

$$c_n = \frac{1}{T} \int_T f(t) e^{-jnwt} \, dt \tag{9.35}$$

Equation 9.34 then becomes

$$f(t) = c_0 + \sum_{n=1}^{\infty} c_n e^{jnwt} + c_{-n} e^{-jnwt} \tag{9.36}$$

If the summation for n goes from $-\infty$ to ∞, this then gives the most compact form for the Fourier series as

$$f(t) = \sum_{n=-\infty}^{\infty} c_n e^{jnwt} \tag{9.37}$$

where,

$$c_n = \frac{1}{T} \int_T f(t) e^{-jnwt} \, dt \tag{9.38}$$

Again, this result for c_n follows directly from the orthogonality condition for exponential notation, that is, from Equation 9.7.

Note: The real coefficients a_n and b_n can be obtained from Equation 9.32 using $c_n = \frac{1}{2}\left(a_n - jb_n\right)$ and $c_{-n} = \frac{1}{2}\left(a_n + jb_n\right)$.

Solving these for a_n and b_n gives

$$a_n = c_n + c_{-n} \quad \text{and} \quad jb_n = c_{-n} - c_n \tag{9.39}$$

These allow for converting back to the real form of the Fourier series if required.
    The next example shows how to find the Fourier series of a periodic function using the complex form.

Example 9.2

Determine the complex Fourier series for the following periodic function:

$$f(t) = t \qquad -\pi < t < \pi \tag{9.40}$$
$$f(t) = f(t + 2\pi)$$

Solution: Sketch the function $f(t)$ as shown in Figure 9.7.
    Using Equations 9.37 and 9.38 for the complex Fourier series,

$$f(t) = \sum_{n=-\infty}^{\infty} c_n e^{jnwt} \tag{9.41}$$

$$c_n = \frac{1}{T} \int_T f(t) e^{-jnwt} \, dt \tag{9.42}$$

Since $f(t) = f(t + 2\pi)$ gives $T = 2\pi$ and so $w = \dfrac{2\pi}{T} = 1$.
Calculating the coefficients c_n using

$$c_n = \frac{1}{T} \int_T f(t) e^{-jnwt} \, dt \tag{9.43}$$

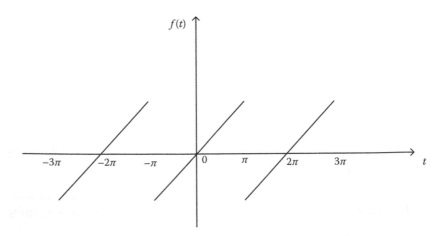

Figure 9.7 Graph of the function $f(t)$.

$$= \frac{1}{2\pi} \int\limits_{-\pi}^{\pi} t\, e^{-jnt}\, dt \qquad (9.44)$$

Using integration by parts to solve Equation 9.44

$$u = t \quad \text{and} \quad \frac{dv}{dt} = e^{-jnt}$$

gives

$$du = dt \quad \text{and} \quad v = \frac{e^{-jnt}}{-jn}$$

And using the formula for integration by parts gives

$$c_n = \frac{1}{2\pi} \left[\left[\frac{t\, e^{-jnt}}{-jn} \right]_{-\pi}^{\pi} - \int\limits_{-\pi}^{\pi} \frac{e^{-jnt}}{-jn}\, dt \right]$$

$$c_n = \frac{1}{2\pi} \left[\frac{-\pi}{jn} \left(\pi\, e^{-jn\pi} - (-\pi) e^{jn\pi} \right) - \left[\frac{e^{-jnt}}{(-jn)^2} \right]_{-\pi}^{\pi} \right]$$

$$c_n = \frac{1}{2\pi} \left[\frac{-\pi}{jn} \left(e^{-jn\pi} + e^{jn\pi} \right) + \frac{1}{n^2} \left(e^{-jn\pi} - e^{jn\pi} \right) \right]$$

Since, $e^{-jn\pi} = e^{jn\pi} = \cos n\pi = (-1)^n$

$$c_n = \frac{1}{2\pi} \left[\frac{-\pi}{jn}\, 2(-1)^n \right]$$

This gives

$$c_n = \frac{j(-1)^n}{n}$$

Therefore, the complex Fourier series becomes

$$f(t) = \sum_{n=-\infty}^{\infty} \frac{j(-1)^n}{n}\, e^{jnwt} \qquad (9.45)$$

Converting back to the real form of the Fourier series can be done using Equation 9.39, that is, $a_n = c_n + c_{-n}$ and $jb_n = c_{-n} - c_n$ if required.

9.4   Fourier Transforms

So far, the Fourier series has been representing periodic functions by a combination of infinite sinusoidal functions. What happens when the function is not periodic in nature? Can it still be made up of a combination of simpler functions?

The idea is to transfer from periodic phenomena to nonperiodic phenomena. This can be achieved by viewing a nonperiodic function as a limiting case of a periodic function as the period tends to infinity.

9.4.1   Nonperiodic Functions

As stated earlier, Fourier series are applicable to periodic functions only, but nonperiodic functions can also be decomposed into Fourier components. This process is called a Fourier transform of a function or signal.

First consider a function that is of finite extent but much less than its periodicity, T, as shown in Figure 9.8. If the period T becomes very large, that is, it tends to infinity, then the above function shown in Figure 9.8 becomes an isolated aperiodic function as required. This limiting process is used to develop a heuristic approach to finding the equations for the Fourier transform from the Fourier series.

9.4.2   Fourier Transform Pair

Starting with the Fourier series and the Fourier coefficient formulae 9.41 and 9.42

$$f(t) = \sum_{n=-\infty}^{\infty} c_n e^{jnwt} \qquad (9.46)$$

$$c_n = \frac{1}{T} \int_T f(t) e^{-jnwt}\, dt \qquad (9.47)$$

It would be nice to just let $T \to \infty$ for c_n in Equation 9.47, but this will not work since it can be shown that $C_n \to 0$ in this case.

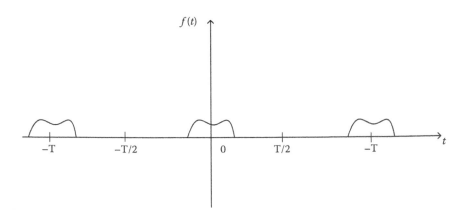

Figure 9.8 A finite function with a large period.

What is considered first is to multiply the equation for c_n by the period T and look at the integral given by

$$\int_T f(t)e^{-jnwt}\,dt \tag{9.48}$$

The period can be any period, so consider going from $-\dfrac{T}{2}$ to $\dfrac{T}{2}$ and replacing for $w = \dfrac{2\pi}{T}$ into Equation 9.48 giving

$$F\left(\frac{n}{T}\right) = \int_{-\frac{T}{2}}^{\frac{T}{2}} f(t)e^{-2\pi j\left(\frac{n}{T}\right)t}\,dt \tag{9.49}$$

Then Equation 9.46 for the Fourier series can now be written as

$$f(t) = \sum_{n=-\infty}^{\infty} F\left(\frac{n}{T}\right)e^{2\pi j\left(\frac{n}{T}\right)t}\frac{1}{T} \tag{9.50}$$

Now as $T \rightarrow \infty$, the "discrete variable" $\dfrac{n}{T}$ gets closer together and is replaced by the continuous variable s, where $-\infty < s < \infty$.

Therefore, Equation 9.49 can now be called the Fourier transform and can be written as

$$F(s) = \int_{-\infty}^{\infty} f(t)e^{-2\pi jst}\,dt \tag{9.51}$$

And the Fourier series Equation 9.49 is given by the summation changing to an integral, and the $\dfrac{1}{T}$ gets smaller and smaller $\approx ds$ in integration for the limiting process giving the following:

$$f(t) = \int_{-\infty}^{\infty} F(s)e^{2\pi jst}\,ds \tag{9.52}$$

Therefore, the equations for the Fourier transform for a nonperiodic function are

$$F(s) = \int_{-\infty}^{\infty} f(t)e^{-2\pi jst}\,dt \tag{9.53}$$

$$f(t) = \int_{-\infty}^{\infty} F(s)e^{2\pi jst}\,ds \tag{9.54}$$

Equation 9.53 is called the Fourier transform of $f(t)$ and Equation 9.54 is called the inverse Fourier transform of $F(s)$.

Note: The definitions and notations for the Fourier transform and its inverse are not rigidly fixed; they can vary by factors of 2π or $\sqrt{2\pi}$ in their equations.

It is useful to see how Equation 9.53 can be used to find the Fourier transform of functions, as shown in the next example.

Example 9.3

Find the Fourier transform of the following rectangular function $f(t) = \prod(t)$ given by

$$
\begin{aligned}
f(t) &= 1 & -\frac{1}{2} < t < \frac{1}{2} \\
&= 0 & |t| \geq \frac{1}{2}
\end{aligned}
\tag{9.55}
$$

Solution: A sketch of the rectangular function is shown in Figure 9.9. Using Equation 9.53, the definition of the Fourier transform gives

$$
F(s) = \int_{-\infty}^{\infty} f(t) e^{-2\pi jst} \, dt
$$

$$
= \int_{-\frac{1}{2}}^{\frac{1}{2}} e^{-2\pi jst} \, dt = \left[\frac{e^{-2\pi jst}}{-2\pi js} \right]_{-\frac{1}{2}}^{\frac{1}{2}} = \frac{e^{-\pi js} - e^{\pi js}}{-2\pi js}
$$

$$
= \frac{e^{-\pi js} - e^{\pi js}}{-2\pi js} = \frac{1}{\pi s} \left[\frac{e^{\pi js} - e^{-\pi js}}{2j} \right]
$$

$$
F(s) = \frac{\sin \pi s}{\pi s}
$$

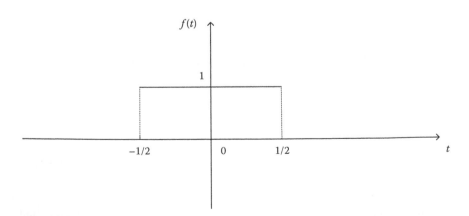

$f(t)$

1

−1/2   0   1/2 t

Figure 9.9 The rectangular function $f(t) = \prod(t)$.

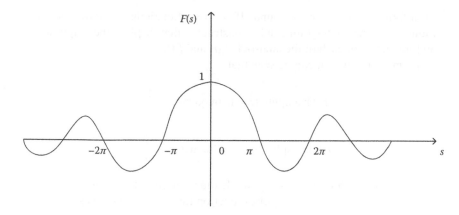

Figure 9.10 The *Sinc(s)* function.

The function $\dfrac{\sin \pi s}{\pi s}$ is given a special name: the *Sinc(s)* function. Graphically, it looks like Figure 9.10.

Fourier transforms of different functions can be found in a similar manner to Example 9.3 but usually there are more involved integration techniques like integration by parts is required (see "Problems" section).

9.4.3 What Does the Fourier Transform Represent?

The essence of the Fourier transform of a waveform is to decompose or separate the waveform into a sum of sinusoids of different frequencies. If these sinusoids sum together to form the original signal waveform, then the Fourier transform of the waveform has been found.

A pictorial representation of the Fourier Transform is a diagram that displays the amplitude and frequency of each of the determined sinusoids as shown in the following example.

Example 9.4

See the general waveform f(t) in Figure 9.11. To find the Fourier transform of $f(t)$ is to ask what combination of sinusoids added together will give $f(t)$. Suppose that $f(t)$ is made up of two functions $f_1(t)$ and $f_2(t)$, then the Fourier

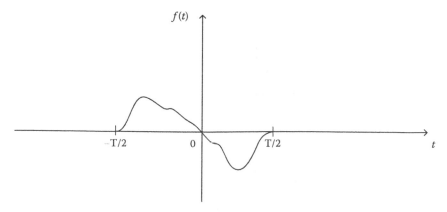

Figure 9.11 A general non-periodic waveform.

Transform of $f(t)$ has been found. If the two functions are shown as in Figure 9.12, then a diagram can be constructed that displays the amplitude and frequency of each of the sinusoids $f_1(t)$ and $f_2(t)$.

From Figure 9.11, it can be seen that

$$f_1(t)\text{: amplitude} = 1;\ \text{frequency} = \frac{1}{T}$$

$$f_2(t)\text{: amplitude} = \frac{1}{2};\ \text{frequency} = \frac{3}{T}$$

Putting this information onto a single diagram gives Figure 9.13. In terms of the delta (or impulse) function this can be written as

$$F(s) = \frac{1}{4}\delta\left(s - \frac{3}{T}\right) + \frac{1}{4}\delta\left(s + \frac{3}{T}\right) - \frac{1}{2}\delta\left(s - \frac{1}{T}\right) - \frac{1}{2}\delta\left(s + \frac{1}{T}\right)$$

A summary of this is to say that every signal has a spectrum and the spectrum then determines the signal.

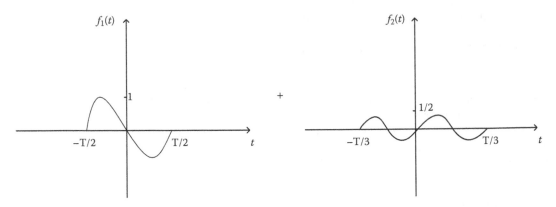

Figure 9.12 Combination of two functions to give $f(t)$.

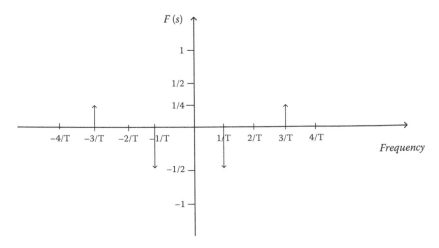

Figure 9.13 Amplitude and frequency components of functions $f_1(t)$ and $f_2(t)$.

9.4.4 Properties of the Fourier Transform

Letting the symbol $\mathcal{F}$ denote the Fourier transform operator, then there are various properties of the Fourier transform similar to the properties of the Laplace transform seen in the last chapter. The first of these is that of the linearity property.

Given a function $f(t)$, then the Fourier transform can be written as

$$\mathcal{F}[f(t)] = F(s)$$

9.4.4.1 Linearity Property

If $f(t)$ and $g(t)$ are functions of time and a and b are constants, then

$$\mathcal{F}\big[a f(t) + b g(t)\big] = a F\big[f(t)\big] + b F\big[g(t)\big]$$
$$= a F(s) + b G(s)$$

There are other important general properties of the Fourier transform and some of these are given in Table 9.1. The proof of the properties can be shown using the definition formula of the Fourier transform given by Equation 9.53. The next example shows the proof for property 3, that is, multiplying a function in the time domain by an exponential produces a shift in the frequency domain.

Table 9.1 Basic Properties of Fourier Transforms

Operation	Time Domain	s-Domain
1. Time shifting	$f(t-a)$	$e^{-2\pi jsa} F(s)$
2. Time scaling	$f(at)$	$\dfrac{1}{|a|} F\left(\dfrac{s}{a}\right)$
3. Multiplying by an exponential in t-domain	$e^{2\pi jat} f(t)$	$F(s-a)$
4. Multiplying by (t)	$t f(t)$	$-\dfrac{1}{2\pi js}\dfrac{d}{ds}[F(s)]$
5. First derivative	$f'(t)$	$2\pi js\, F(s)$
6. Second-order derivative	$f''(t)$	$(2\pi js)^2\, F(s)$
7. nth derivative	$f^n(t)$	$(2\pi js)^n\, F(s)$

Example 9.5

Show that the $\mathcal{F}[e^{2\pi jat} f(t)] = F(s-a)$.

Solution: Starting with the formula for the Fourier transform

$$F(s) = \int_{-\infty}^{\infty} f(t) e^{-2\pi jst}\, dt \qquad (9.56)$$

Substituting for $f(t)$ by $e^{2\pi jat} f(t)$ gives

$$\mathcal{F}\big[e^{2\pi jat} f(t)\big] = \int_{-\infty}^{\infty} e^{2\pi jat} f(t) e^{-2\pi jst}\, dt \qquad (9.57)$$

Now putting the exponentials together gives

$$= \int\limits_{-\infty}^{\infty} f(t)e^{-2\pi j(s-a)t}\, dt = F(s-a) \qquad (9.58)$$

Here, s is being replaced by $s - a$ in the definition of the Fourier transform as required.

This is a very important property and is used in amplitude modulation of signals as shown in the applications section at the end of the chapter.

9.4.5   Convolution of Two Functions

One of the most important operations in signal processing is the idea of the convolution of functions in the time domain. Signal processing uses one signal to modify another. Most often looking to modify the spectrum of a signal this can be achieved at a basic level using the linearity property:

$$\mathcal{F}\big[f(t)+g(t)\big] = \mathcal{F}\big[f(t)\big] + \mathcal{F}\big[g(t)\big]$$

This is just modifying $\mathcal{F}[f(t)]$ by adding the spectrum of $\mathcal{F}[g(t)]$ to it. Therefore, what about multiplying the Fourier transform of two functions does this have a similar correspondence in the time domain of just multiplying the functions? The answer to this is not as simple as just simply multiplication of functions in the time domain but of a convolution of two functions in the time domain instead.

To show this process, start with the multiplication of the Fourier transform of two functions:

$$\mathcal{F}\big[g(t)\big] \times \mathcal{F}\big[f(t)\big] \qquad (9.59)$$

Replacing for the definitions of the Fourier transforms for both $g(t)$ and $f(t)$ gives

$$= \left(\int\limits_{-\infty}^{\infty} g(t)e^{-2\pi jst}\, dt \right)\left(\int\limits_{-\infty}^{\infty} f(x)e^{-2\pi jsx}\, dx \right) \qquad (9.60)$$

This is a separated integral and can be expressed as a mixed double integral as follows:

$$= \int\limits_{-\infty}^{\infty}\int\limits_{-\infty}^{\infty} e^{-2\pi jst}e^{-2\pi jsx}g(t)f(x)\, dt\, dx \qquad (9.61)$$

Collecting the exponential terms together gives

$$= \int\limits_{-\infty}^{\infty}\int\limits_{-\infty}^{\infty} e^{-2\pi js(t+x)}g(t)\, f(x)\, dt\, dx \qquad (9.62)$$

Regrouping the integral as

$$= \int_{-\infty}^{\infty} \left(\int_{-\infty}^{\infty} e^{-2\pi js(t+x)} g(t) dt \right) f(x) dx \tag{9.63}$$

Now, making a change of variables by letting $u = t + x$, $t = u - x$, and $du = dt$, and substituting into Equation 9.63 gives

$$= \int_{-\infty}^{\infty} \left(\int_{-\infty}^{\infty} e^{-2\pi jsu} g(u - x) du \right) f(x) dx \tag{9.64}$$

Interchanging the terms gives

$$= \int_{-\infty}^{\infty} \left(\int_{-\infty}^{\infty} g(u - x) f(x) dx \right) e^{-2\pi jsu} du \tag{9.65}$$

Now defining $h(u)$ as

$$h(u) = \int_{-\infty}^{\infty} g(u - x) f(x) dx \tag{9.66}$$

Then defining the Fourier transform $h(u)$ by

$$\mathcal{F}[h(u)] = \int_{-\infty}^{\infty} h(u) e^{-2\pi jsu} du \tag{9.67}$$

Therefore, it has been shown that

$$\mathcal{F}\left[g(t)\right] \times \mathcal{F}\left[f(t)\right] = \mathcal{F}\left[h(u)\right] \tag{9.68}$$

Defining the convolution of functions g and f as

$$(g * f)x = \int_{-\infty}^{\infty} g(x - y) f(y) dy \tag{9.69}$$

Then finally, it can be stated that the Fourier transform of the convolution of two functions is equal to the product of the Fourier transform of the two functions as shown next:

$$\mathcal{F}\left[g * f\right] = \mathcal{F}\left[g(t)\right] \times \mathcal{F}\left[f(t)\right] \tag{9.70}$$

This is again an important property for signal processing purposes as seen in the applications section.

9.5   Applications

Example 9.6:   Designing of Fourier Transform Infrared (FTIR) Smoke Detectors

Fire detector systems should have the ability to discriminate between real fire sources and nonfire sources. Smoke detectors that can respond quickly may suffer from the inability to discriminate between a real fire smoke and smoke from other sources. In high-value installations where there is expensive and sensitive equipment, it is clear that reliable fire detection systems are needed. The detection systems are generally used to activate fixed fire suppression systems like sprinklers, so any false alarms are an undesirable outcome causing valuable time loss and having potential cost implications.

Fourier transform infrared (FTIR) smoke detectors make use of Fourier Transform techniques to analyze combustion products and so aid new more reliable detection systems to be built. Using FTIR measurements of concentrations of gases (e.g., CO_2, CO, H_2O, and CH_4) given off during different modes of combustion, the detection system can classify input data as a flaming fire, smoldering fire, nuisance, or other environmental sources. Therefore, FTIR spectroscopy can give multiple gas concentrations that enable an advanced fire detection system to be built.

Example 9.7:   Antenna Design Using Frequency Modulation

Consider the design of an antenna for the transmission of an audio signal $x(t)$ with maximum frequency of approximately 10 KHz.

The relationship between the frequency and wavelength is given by the equation

$$\lambda = \frac{c}{f}$$

where λ is the wavelength, f is the frequency, and c is the speed of electromagnetic waves.

Using the values of $f = 10000$ Hz and $c = 3 \times 10^8$ ms^{-1} gives a value for the wavelength of $\lambda = 30,000$ m.

In the design of antennae, they usually have dimensions of $\approx \frac{1}{4}\lambda$. This implies an antenna size of approximately 7500 m, which of course is not practical for design purposes.

One way around this problem is to use frequency modulation and use a carrier signal, say, $\cos 2\pi\omega_c t$, where ω_c is a very high frequency combined with the original signal. The new modulated function can be written as $\phi(t) = x(t) \cos 2\pi\omega_c t$.

Using the definition of the exponential form for the cosine function gives

$$\phi(t) = x(t)\frac{1}{2}\left(e^{2\pi j\omega_c t} + e^{-2\pi j\omega_c t}\right)$$

If the Fourier transform of $x(t)$ is $X(s)$ and the Fourier Transform of $\phi(t)$ is $\Phi(s)$ then making use of property 3 in Table 9.1 gives

$$\Phi(s) = \frac{1}{2}\left[X(s - \omega_c) + X(s + \omega_c)\right]$$

Graphically, this is shown in Figure 9.14.

Clearly, the frequency for the modulated signal is now higher than the original signal, which then makes for a smaller λ and so making for a more realistic smaller antenna size.

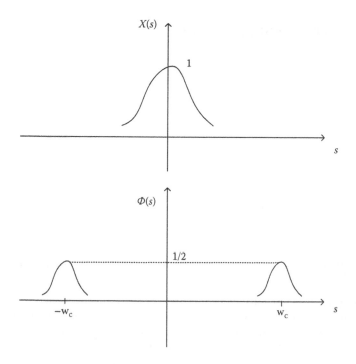

Figure 9.14 Frequency for the modulated signal $\Phi(s)$.

Example 9.8: Smoothing Process Using Filters

In Section 9.4.5, it was found that the Fourier transform of the convolution of two functions g and f is given by the product of the Fourier transforms of the individual functions. This property can be used in signal processing as a method of *filtering* the required signal.

Taking as an example a set of results from an experiment. The results have been plotted on a graph as a function (or signal) $\phi(t)$ that is periodic in nature but has a jagged appearance around the edges (see Figure 9.15).

Taking the Fourier transform of this signal will show the spectrum of the frequencies associated with the signal $\Phi(s)$. This is shown in Figure 9.16.

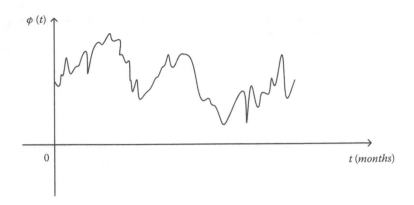

Figure 9.15 Signal showing jagged edges.

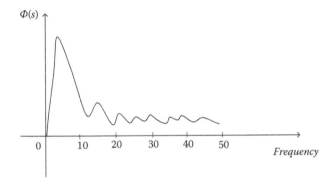

Figure 9.16 Graph showing the frequency spectrum of the signal $\Phi(s)$.

Now the high-end frequencies tend to cause the jaggedness or rapid oscillations with the signal. If the high frequencies could be eliminated somehow, then the signal may become smoother in appearance.

This can be achieved by multiplying the signal $\Phi(s)$ by a scaled rectangular function $\Pi_{2\nu}(s)$ in the frequency domain as shown in Figure 9.17.

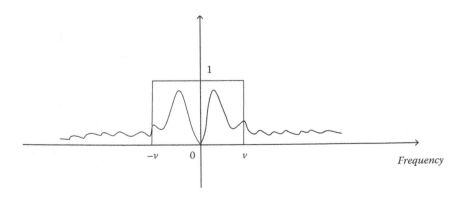

Figure 9.17 Multiplying $\Phi(s)$ by the rectangular function $\Pi_{2\nu}(s)$.

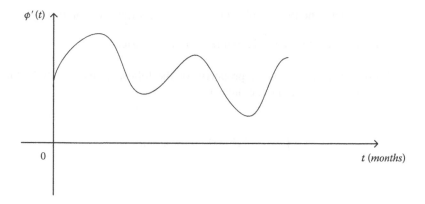

Figure 9.18 The convoluted function $\phi'(t)$.

This is called a low-pass filter as it allows the lower frequencies through and cuts out the higher frequencies. In the frequency domain this is given by multiplying the two functions, that is, $\Pi_{2v}(s) \times \Phi(s)$ and in the time domain, this will correspond to a convolution of the *Sinc* function with $\phi(t)$ to produce $\phi'(t)$, that is,

$$\phi'(t) = 2v \; Sinc \; (2vt)* \; \phi(t)$$

The result $\phi'(t)$ is a much smoother function with the jaggedness taken out, as shown in Figure 9.18.

Examples can also be found where the use of other types of filtering, such as, high-pass filters and band-pass filters that allow signals to be manipulated to produce the required outputs.

Problems

9.1 Determine the trigonometric Fourier series for the following functions:

a. $f(t) = 0 \qquad\qquad -2 < t < 0$
$\qquad = t \qquad\qquad\quad 0 < t < 2$
$\quad f(t) = f(t+4)$

b. $g(t) = 1 \qquad\qquad 0 \leq t < \pi$
$\qquad = 0 \qquad\qquad -\pi \leq t < 0$
$\quad g(t) = g(t + 2\pi)$

9.2 Given the following periodic function $h(t)$:

$$h(t) = t - 3 \qquad -2 < t < 0$$
$$= t + 3 \qquad\quad 0 < t < 2$$
$$h(t) = h(t + 4)$$

    a. Determine the complex Fourier series for the function $h(t)$.

    b. Hence, obtain its trigonometric Fourier series.

9.3 Prove the orthogonality property for the following exponential functions, where m and n are integers.

$$\int_{-\pi}^{\pi} e^{jnt} e^{-jmt}\, dt = 0 \qquad m \neq n$$

$$= 2\pi \qquad m = n$$

9.4 Given the triangular function $\Lambda(t)$ is shown in Figure 9.19 and defined below as

$$\Lambda(t) = 1 - |t| \qquad |t| \leq 1$$
$$= 0 \qquad |t| > 1$$

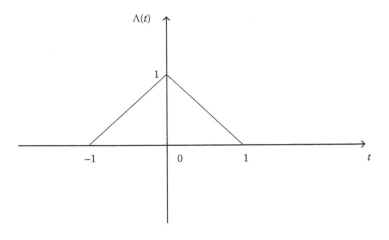

Figure 9.19 The triangular function $\Lambda(t)$.

show that the Fourier transform is given by $\mathcal{F}[\Lambda(t)] = Sinc^2\, s$.

9.5 Given the Gaussian function $f(t) = e^{-\pi t^2}$, show that its Fourier transform is given by $F(s) = e^{-\pi s^2}$. (Harder problem)

(Hint: Start with definition, consider $F'(s)$ and then integration by parts.)

10 Multivariable Calculus

Multivariable calculus is essentially an extension of the calculus with one variable in which the system now depends on many variables. In engineering, multivariable calculus can be used to model higher dimensional system behavior, that is, stress may depend on the x, y, and z positions. Most of the concepts introduced in the one-variable calculus discussions can be extended to multivariable calculus starting with the ideas of partial derivatives. The higher-order partial derivatives lead onto the multivariable chain rule and ideas of a general directional derivative with applications to tangent planes. Higher-order integration of double and triple integrals using different coordinate systems is considered as well as how these concepts are used in real-world applications.

10.1 Partial Derivatives

10.1.1 Introduction and Definition

A function of two or more variables such as

$$f(x, y) = xy - ye^x$$

has two inputs x and y and it produces one output. A many-input function could have many outputs and is usually termed as a vector function.

Recall the definition from 1-dimensional differentiation (or derivative) as seen in Chapter 6,

$$\frac{df(x)}{dx} = f'(x) = \lim_{h \to 0} \left[\frac{f(x+h) - f(x)}{h} \right] \tag{10.1}$$

What is the interpretation of derivatives in 1-dimension calculus? The first is to consider it as the slope of the tangent line as shown in Figure 10.1.

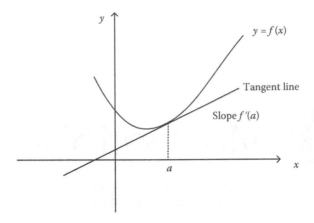

Figure 10.1 The derivative as the slope of the tangent line at some point.

Second, the derivative can be represented as the instantaneous rate of change of a function, that is, using the first two terms of a Taylor series gives

$$f(a+h) \approx f(a) + hf'(a) \tag{10.2}$$

If there is a change in the input by an amount h, the output changes by the rate of change output $f'(a)$ multiplied by h. So $f'(a)$ represents the instantaneous rate of change of $f(a)$.

Now what is the situation when there is two or more variables? Here there is still a definition of derivatives, but since there is more than one variable these are called *partial derivatives*. Considering a function with two variables, the partial derivatives can be defined in a similar manner to Equation 10.1.

10.1.1.1 Partial Derivatives Defined

A function $f(x, y)$ has two partial derivatives:

Partial with respect to x: $\quad \dfrac{\partial f}{\partial x}(a,b) = f_x \triangleq \lim_{h \to 0} \left[\dfrac{f(a+h,b) - f(a,b)}{h} \right]$

Partial with respect to y: $\quad \dfrac{\partial f}{\partial y}(a,b) = f_y \triangleq \lim_{h \to 0} \left[\dfrac{f(a,b+h) - f(a,b)}{h} \right]$

So, with partial derivatives the derivatives are taken with respect to one variable while keeping the other variables fixed.

Note: The shorthand notation for partial derivatives is $\dfrac{\partial f}{\partial x}(a,b) = f_x.$

In the same way as with 1-D derivatives, the interpretation of the partial derivatives can be seen as an instantaneous rate of change as follows.

10.1.1.2   *Instantaneous Rate of Change*

$$f(a+h,b) \approx f(a,b) + h\frac{\partial f}{\partial x}(a,b)$$

$$f(a,b+h) \approx f(a,b) + h\frac{\partial f}{\partial y}(a,b)$$

These can be put together as a single total change given by

$$f(a+h,b+k) \approx f(a,b) + h\frac{\partial f}{\partial x}(a,b) + k\frac{\partial f}{\partial y}(a,b) \qquad (10.3)$$

Here a small change in the input variables a by h and b by k produces a total output change given by Equation 10.3. Also, this can be thought of as the slope of a tangent line to a surface as shown in Figure 10.2.

Another important way in considering derivatives in higher dimensions is to fix all the inputs except one and think of it as a function of the one variable only. For example,

$$g(x) = f(x,b) \text{ with } y = b$$

then

$$g'(a) = \frac{\partial f}{\partial x}(a,b)$$

at the point $x = a$.

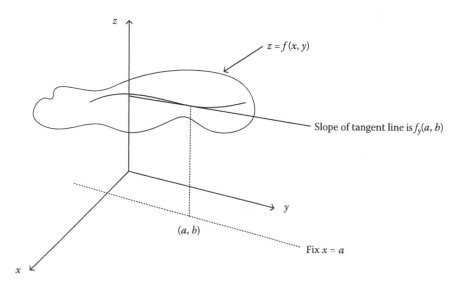

Figure 10.2 Shows the slope of the tangent line keeping one variable fixed.

This is generally the best way to understand partial derivatives and is the way that is used to compute them. The next examples illustrate this process.

Example 10.1

Given $f(x, y) = x + xy^2$, compute the following.

1. $\dfrac{\partial f}{\partial x}(1,3)$

2. $\dfrac{\partial f}{\partial y}(2,4)$

Solution:

1. For $\dfrac{\partial f}{\partial x}(1,3)$, here x is varying and so consider y as fixed or constant.

   $f(x, y) = x + xy^2$ differentiating with respect to x keeping y fixed

   gives $\dfrac{\partial f}{\partial x} = 1 + y^2$.

   Now substituting in the values of the point (1,3) gives

$$\frac{\partial f}{\partial x}(1,3) = 1 + 3^2 = 10$$

2. For $\dfrac{\partial f}{\partial y}(2,4)$, here y is varying and consider x as fixed or constant.

   $\dfrac{\partial f}{\partial y} = 2xy$, so $\dfrac{\partial f}{\partial y}(2,4) = 2(2)(4) = 16$.

Example 10.2

Given $f(x, y) = x^2y + xe^y$,   compute   (1) $f_x(1, 0)$ and   (2) $f_y(1, 1)$.

Solution:

1. $f_x = \dfrac{\partial f}{\partial x} = 2xy + e^y$

$$f_x(1,0) = 2(1)(0) + e^0 = 1$$

2. $f_y = \dfrac{\partial f}{\partial y} = x^2 + xe^y$

$$f_y(1,1) = (1)^2 + (1)e^1 = 1 + e$$

10.1.2   Higher Derivatives

As with the one variable case higher derivatives can be calculated. The only difference now is that with the multiple variable case, there are many higher derivatives to consider. With a single variable case, there is only one second derivative but with a two-variable problem. It turns out that there are four second derivatives that can be calculated. Some of the notation used is given next.

The following notation is used:

$$\frac{\partial^2 f}{\partial x^2} = \frac{\partial}{\partial x}\left[\frac{\partial f}{\partial x}\right] = f_{xx}$$

$$\frac{\partial^2 f}{\partial y^2} = \frac{\partial}{\partial y}\left[\frac{\partial f}{\partial y}\right] = f_{yy}$$

$$\frac{\partial^2 f}{\partial y \partial x} = \frac{\partial}{\partial y}\left[\frac{\partial f}{\partial x}\right] = f_{xy}$$

$$\frac{\partial^2 f}{\partial x \partial y} = \frac{\partial}{\partial x}\left[\frac{\partial f}{\partial y}\right] = f_{yx}$$

Example 10.3

Given $f(x,y) = xe^y + xy^3$, find the following second-order derivatives:

(1) f_{xx} (2) f_{xy} (3) f_{yx} (4) f_{yy}.

Solution:

1. $f_x = e^y + y^3$, $f_{xx} = 0$
2. $f_x = e^y + y^3$, $f_{xy} = e^y + 3y^2$
3. $f_y = xe^y + 3xy^2$, $f_{yx} = e^y + 3y^2$
4. $f_y = xe^y + 3xy^2$, $f_{yy} = xe^y + 6xy$

It can be seen that $f_{xy} = f_{yx}$. Generally, this property is observed and the following theorem known as Clairaut's theorem expresses this.

10.1.2.1   Clairaut's Theorem

Most of the time $f_{xy}(a, b) = f_{yx}(a, b)$. The order of the differentiation does not matter. With more variables, that is, x, y, z, the order does not matter, for example, $f_{xyzz} = f_{zxzy}$. However, there are conditions on f for this to be true.

10.1.2.2   Antiderivatives When There Are Multiple Variables

Example 10.4

Find a function $f(x,y)$ such that

$$f_x = 2x + 3y$$

$$f_y = 3x + e^y$$

Solution: You need to find the antiderivative of f_x by keeping y constant. So, starting with $f_x = 2x + 3y$ and integrating with respect to x keeping y as a constant implies

$$f = x^2 + 3xy + g(y) \text{ (any function of } y)\tag{10.4}$$

Finding the antiderivative of f_y keeping x constant implies

$$f = 3xy + e^y + h(x) \text{ (any function of } x)\tag{10.5}$$

Now the function $f(x, y)$ has to fit both Equation 10.4 and Equation 10.5. Therefore,

$$f(x,y) = x^2 + 3xy + e^y$$

will satisfy both the required conditions. This will be an important idea for working with conservative vector fields as seen later in Chapter 11.

Example 10.5

Find $f(x, y)$ such that

$$f_x = 4x - 3y$$

$$f_y = x + y$$

Solution:

$$f_x = 4x - 3y \quad \text{implies} \quad f = 2x^2 - 3xy + g(y)$$

$$f_y = x + y \quad \text{implies} \quad f = xy + \frac{y^2}{2} + h(x)$$

Now in this case there cannot be found any function $f(x, y)$ that fits both profiles. Therefore, it is not possible to find any $f(x, y)$ that has these required f_x and f_y functions simultaneously.

10.1.3 Chain Rule

10.1.3.1 Chain Rule with One Variable

The chain rule in one variable is a process of calculating derivatives in situations where there is a function of a function involved. This can be stated in different ways, including

$$\frac{d}{dx}\left[f\left[g(x)\right]\right] = f'\left[g(x)\right]g'(x)\tag{10.6}$$

But this can also be thought of as

$$f = f(y) \text{ and } g = g(x)$$

$$\frac{df}{dx} = \frac{df}{dy} \times \frac{dy}{dx} \tag{10.7}$$

Sometimes it is easier to see how the variables are related to each other by drawing a dependency chart as shown in Figure 10.3. Equations 10.6 and 10.7 are the same process.

Using Equation 10.7 as the way of thinking about how variables depend on each other is more useful as it leads to a natural extension in higher dimensions as shown in the next sections.

Figure 10.3 Dependency chart showing relationship between variables.

10.1.3.2 Chain Rule with Multivariables

Suppose there is a function $f(x, y)$ and

$$x = x(u, v)$$

$$y = y(u, v)$$

Then what is the change in the function due to a change in the variable u, that is, $\frac{\partial f}{\partial u}$?

First, drawing a dependency chart helps to show the situation of how the variables are related to each other, as in Figure 10.4.

So,

$$\frac{\partial f}{\partial u} = \frac{\partial f}{\partial x} \times \frac{\partial x}{\partial u} + \frac{\partial f}{\partial y} \times \frac{\partial y}{\partial u}$$

This gives the net chain in f due to the change u through the variables x and y.

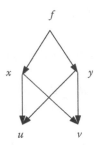

Figure 10.4 Dependency chart showing relationship between variables.

Example 10.6

Given that $f = f(z,w)$, $z = z(x,y)$, $w = w(y)$, $x = x(u)$, and $y = y(u,v)$, find $\dfrac{\partial f}{\partial u}$ and $\dfrac{\partial f}{\partial v}$.

Solution: First draw a dependency chart showing the relationship between the variables as in Figure 10.5.

So, the change in f due to a change in u is given as

$$\frac{\partial f}{\partial u} = \frac{\partial f}{\partial z}\frac{\partial z}{\partial x}\frac{\partial x}{\partial u} + \frac{\partial f}{\partial w}\frac{\partial w}{\partial y}\frac{\partial y}{\partial u} + \frac{\partial f}{\partial z}\frac{\partial z}{\partial y}\frac{\partial y}{\partial u}$$

Also, the change in f due to a change in v is given by

$$\frac{\partial f}{\partial v} = \frac{\partial f}{\partial z}\frac{\partial z}{\partial y}\frac{\partial y}{\partial v} + \frac{\partial f}{\partial w}\frac{\partial w}{\partial y}\frac{\partial y}{\partial v}$$

Consider the chain rule by using the diagram to find the relationships between the different variables.

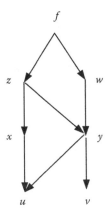

Figure 10.5 Dependency chart showing relationships between variables.

Example 10.7

Given that $f = f(u,v,t)$ where $u = u(t)$ and $v = v(t)$, find $\dfrac{\partial f}{\partial t}$.

Solution: First draw the dependency chart as in Figure 10.6.

$$\frac{df}{dt} = \frac{\partial f}{\partial u}\frac{\partial u}{\partial t} + \frac{\partial f}{\partial v}\frac{\partial v}{\partial t} + \frac{\partial f}{\partial t}$$

Total derivative
(with the variable t)

Partial derivative, how f changes with
the third variable input t

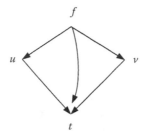

Figure 10.6 Dependency chart showing relationships between variables.

The next section considers derivatives in a more general direction called the directional derivative.

10.1.4   Directional Derivatives and Gradients

To find the change of a function $f(x,y)$ in any general direction not just the x and y directions (i.e., the $\dfrac{\partial f}{\partial x}$ and $\dfrac{\partial f}{\partial y}$), it is important to consider the following situation shown in Figure 10.7. What is the directional derivative of $f(x,y)$ at (x_0,y_0) in the direction of $\langle a, b\rangle$?

Parameterization of the line in the direction $\langle a, b\rangle$ is given by

$$x = x_0 + \frac{a}{\sqrt{a^2 + b^2}}t \tag{10.8}$$

$$y = y_0 + \frac{b}{\sqrt{a^2 + b^2}}t \tag{10.9}$$

These makes us travel along $\langle a, b\rangle$ with unit speed so unit time is now equal to unit distance in the direction $\langle a, b\rangle$. A dependency chart for this situation is shown in Figure 10.8.

Now the change in f due to a change in t is given by

$$\frac{\partial f}{\partial t} = \frac{\partial f}{\partial x}\frac{dx}{dt} + \frac{\partial f}{\partial y}\frac{dy}{dt}$$

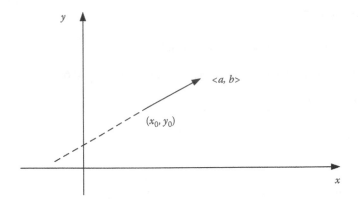

Figure 10.7 General directional vector.

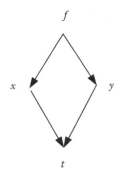

Figure 10.8 Dependency chart showing relationships between variables.

Using Equations 10.8 and 10.9 and differentiating with respect to t gives

$$\frac{\partial f}{\partial t} = \frac{\partial f}{\partial x}\frac{a}{\sqrt{a^2+b^2}} + \frac{\partial f}{\partial y}\frac{b}{\sqrt{a^2+b^2}} \tag{10.10}$$

To represent this change of f with a change in t in the direction of $\langle a,b\rangle$ as a special notation it is written as $D_{\langle a,b\rangle}f$, the directional derivative of f in the direction of $\langle a,b\rangle$. Equation 10.10 can now be written in a more compact form using the dot product form as

$$D_{\langle a,b\rangle}f = \langle\frac{\partial f}{\partial x},\frac{\partial f}{\partial y}\rangle \cdot \frac{\langle a,b\rangle}{\sqrt{a^2+b^2}} \tag{10.11}$$

Using some further notation as $\langle\frac{\partial f}{\partial x},\frac{\partial f}{\partial y}\rangle = \nabla f$ (called the gradient of f). Also, ∇ is sometimes called "del" or "nabla." So, Equation 10.11 can be written in simpler notation as follows.

Given $f(x, y)$ with its gradient $\nabla f = \langle f_x, f_y \rangle$, the directional derivative of f in the direction $\bar{u}$ is given by

$$D_{\bar{u}}f = \nabla f \cdot \frac{\bar{u}}{|\bar{u}|} = \nabla f \cdot \hat{u} \qquad (10.12)$$

Now it can be checked to see what the directional derivatives would be in the x and y directions from the formula given in Equation 10.12 as follows.

In the x-direction, the unit vector is $\hat{u} = \langle 1, 0 \rangle$. This then gives $D_{\bar{u}}f$ as

$$D_{\langle 1, 0 \rangle}f = \nabla f . \langle 1, 0 \rangle = \langle f_x, f_y \rangle . \langle 1, 0 \rangle = f_x$$

and in the y-direction $D_{\bar{u}}f$ is

$$D_{\langle 0, 1 \rangle}f = \nabla f . \langle 0, 1 \rangle = \langle f_x, f_y \rangle . \langle 0, 1 \rangle = f_y$$

as expected for both.

Example 10.8

Given $f(x,y) = xy^2 - 10x$:

1. Compute ∇f.
2. What is the directional derivative of f in the direction $\langle 2,5 \rangle$ at the point $(1, 1)$.

Solution:

1. $\nabla f = \langle f_x, f_y \rangle = \langle y^2 - 10, 2xy \rangle$

2. $D_{\langle 2,5 \rangle} f(1,1) = \nabla f(1,1). \dfrac{\langle 2,5 \rangle}{\sqrt{29}}$

$$= \langle -9, 2 \rangle \cdot \frac{\langle 2,5 \rangle}{\sqrt{29}} = -\frac{8}{\sqrt{29}}$$

Example 10.9

Given $f(x,y) = x - xy^2$, find:

1. $D_{\langle 2,-1 \rangle} f(1,0)$
2. In what direction $\hat{u}$ is $D_{\bar{u}}f$ the biggest?

Solution:

1. Now $\nabla f = \langle f_x, f_y \rangle = \langle 1 - y^2, -2xy \rangle$.
   So, $\nabla f(1,0) = \langle 1,0 \rangle$.

Therefore,

$$D_{\langle 2,-1 \rangle}f(1,0) = \langle 1,0 \rangle \cdot \frac{\langle 2,-1 \rangle}{\sqrt{5}} = \frac{2}{\sqrt{5}}$$

2.

$$D_{\hat{u}}f(1,0) = \nabla f(1,0).\hat{u}$$
$$= |\nabla f||\hat{u}|\cos\theta$$

To make this the biggest, $\cos\theta$ needs to be the largest, which implies $\theta = 0$. Therefore, angle between ∇f and $\hat{u}$ should be in the same direction. So, $\hat{u}$ should be in the same direction as ∇f but the unit vector version of it, that is,

$$\hat{u} = \frac{\nabla f}{|\nabla f|}$$

For the above problem $\nabla f = \langle 1,0 \rangle$, therefore $\hat{u} = \langle 1,0 \rangle$. So, $\nabla f = \langle f_x, f_y \rangle$ always points in the direction of steepest increase.

10.1.5 Stationary Points (Maxima, Minima, and Saddle Points)

To try to find the maximum and minimum of the function of two variables, let's first review the theory in the one-variable problem again as shown in Figure 10.9.

The tangent line gives a first-order approximation:

$$f(x) \approx f(a) + (x-a)f'(a)$$

And if $y \approx f(a) + (x-a)f'(a)$ this is just the tangent line, similar to the arguments for the tangent plane in higher-order problems.

What do you consider if it is a maximum? At a maximum, $f'(a) = 0$. So, the second-order approximation is shown in Figure 10.10.

$$f(x) \approx f(a) + (x-a)f'(a) + \frac{1}{2}(x-a)^2 f''(a)$$

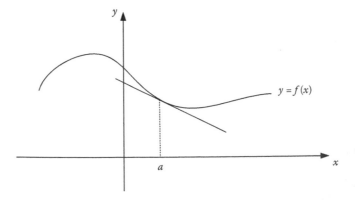

Figure 10.9 Tangent line at a point.

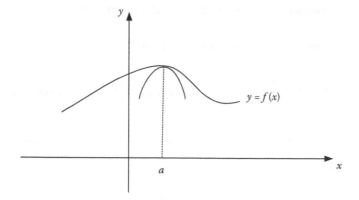

Figure 10.10 Second-order approximation at given point.

This is the Taylor series expansion of a function at a point. This is the parabola of best fit at $x = a$. Now, if $f'(a) = 0$ at a maximum, then $f(x)$ becomes

$$f(x) \approx f(a) + \frac{1}{2}(x-a)^2 f''(a)$$

So, this parabola is $\smile$ shaped if $f''(a) > 0$ (i.e., a minimum) and this parabola is $\wedge$ shaped if $f''(a) < 0$ (i.e., a maximum).

This can be extended to a function of higher variables. Using a Taylor series expansion of a function of two variables gives

$$f(x,y) = f(a,b) + \nabla f(a,b).\langle x-a, y-b \rangle + \frac{1}{2}\langle x-a, y-b \rangle.\nabla^2 f(a, b).\langle x-a, y-b \rangle$$

There needs to be $\nabla f(a,b) = \bar{0}$ for maximum and minimum. And

$$\nabla^2 f = \begin{pmatrix} f_{xx} & f_{xy} \\ f_{yx} & f_{yy} \end{pmatrix}$$

is a matrix called the "Hessian."

Note: The determinant of the above matrix is also sometimes referred to as the Hessian.

To determine if you have a maximum or minimum you need to consider the Hessian matrix. If D is called the Hessian, then it is defined as

$$D = \det \begin{pmatrix} f_{xx} & f_{xy} \\ f_{yx} & f_{yy} \end{pmatrix}$$

and it is this that determines whether there is a maximum or minimum.
    Conditions for determining maximum or minimum are stated next.

10.1.5.1   Summary to Find Maximum or Minimum Points

1. Find the critical points given by $\nabla f(a,b) = \overline{0}$.

2. Let $D = \det \begin{pmatrix} f_{xx} & f_{xy} \\ f_{yx} & f_{yy} \end{pmatrix} = f_{xx}(a,b) f_{yy}(a,b) - f_{xy}^2 = $ Discriminant

If $D > 0$ and $f_{xx}(a,b) \; > \; 0$,   then (a,b) is a local minimum.

If $D > 0$ and $f_{xx}(a,b) \; < \; 0$,   then (a,b) is a local maximum.

If $D < 0$,   then (a,b) is a saddle point.

Example 10.10

Find and identify the critical points of the following function $f(x, y)$:

$$f(x,y) = x^3 - 12xy + 8y^3$$

Solution:

$$0 = f_x = 3x^2 - 12y \;\; \text{gives} \;\; x^2 = 4y$$

$$0 = f_y = -12x + 24y^2 \;\; \text{gives} \;\; x = 2y^2$$

Using $x = 2y^2$ into $x^2 = 4y$ gives $(2y^2)^2 = 4y$. Solving for y gives $4y(y^3 - 1) = 0$. This gives $y = 0$ or $y = 1$ as the solutions to this equation which, then gives the critical points as $(0,0)$ and $(2,1)$.

To see if the critical points are a maximum or minimum, find the Hessian D:

$$D = \det \begin{pmatrix} f_{xx} & f_{xy} \\ f_{yx} & f_{yy} \end{pmatrix} = f_{xx}(a,b) f_{yy}(a,b) - f_{xy}^2$$

Calculating all the higher-order derivatives gives

$$f_{xx} = 6x \quad f_{xx}(0,0) = 0 \quad f_{xx}(2,1) = 12$$

$$f_{yy} = 48y \quad f_{yy}(0,0) = 0 \quad f_{yy}(2,1) = 48$$

$$f_{xy} = -12 \quad f_{xy}(0,0) = -12 \quad f_{xy}(2,1) = -12$$

So, $D(0,0) = (0)(0) - 144$ gives $D = -144 < 0$, which implies a saddle point.

$D(2,1) = (12)(48) - 144$ gives $D = 432 > 0$, but $f_{xx}(2,1) > 0$, so it is a minimum.

Example 10.11

Find three numbers that sum to 100 and have the largest product.

Solution: Let x, y, and $100 - x - y$ be the three numbers. Therefore, maximize $f(x,y) = xy(100 - x - y)$.
    Look for critical points:

$$0 = f_x = y(100 - x - y) + x(-y) = y(100 - 2x - y)$$

$$0 = f_y = x(100 - x - y) + y(-x) = x(100 - x - 2y)$$

Since the product of two things equals zero implies $y = 0$, the product $= 0$, so ignore this.
    Therefore, $100 - 2x - y = 0$
    and $x = 0$, then again the product $= 0$, so ignore this.
    Therefore, $100 - x - 2y = 0$.
    Solving these two equations simultaneously gives

$$2x + y = 100$$

$$x + 2y = 100$$

$$-3y = -100 \quad y = \frac{100}{3}$$

This gives $x = \frac{100}{3}$ and the third number $100 - x - y = \frac{100}{3}$.
    So the numbers are $\left(\frac{100}{3}, \frac{100}{3}, \frac{100}{3} \right)$.

10.2   Higher-Order Integration

10.2.1   Double Integrals and Fubini's Theorem

Instead of integrating over an interval, integration is carried out as shown in Figure 10.11 over a 2-D region.
    First, chop the region into small pieces and pick a point (x^*, y^*). Then,

$$\iint\limits_{D} f(x, y) \, dA \triangleq \lim_{\Delta A \to 0} \sum_{all \; pieces} f(x^*, y^*) \Delta A$$

To compute this double integral, start with a special case of a rectangle as shown in Figure 10.12.

Note: *The integration sign is used twice.*

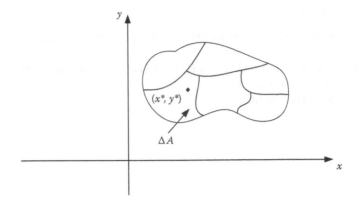

Figure 10.11 A general 2-D region.

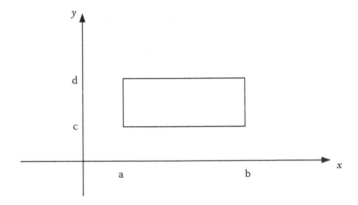

Figure 10.12 A general rectangular area.

$$\iint\limits_D f(x,y)\,dA \triangleq \lim_{\Delta A \to 0} \sum_{all\ pieces} f(x^*,y^*)\Delta A$$

Split the rectangular region into horizontal and vertical strips as shown in Figure 10.13.

$$\lim_{\Delta x \to 0}\lim_{\Delta y \to 0}\sum_i\sum_j f(x^*,y^*)\Delta y \Delta x$$

$$= \lim_{\Delta x \to 0}\sum_i\left[\sum_j \lim_{\Delta y \to 0} f(x^*,\ y^*)\Delta y\right]\Delta x$$

$$= \lim_{\Delta x \to 0}\sum_i\int_c^d f(x^*,y)\,dy\Delta x$$

$$= \int_c^d \lim_{\Delta x \to 0}\sum_i f(x^*,y)dy\Delta x$$

$$\int_a^b\int_c^d f(x,y)\,dy\,dx$$

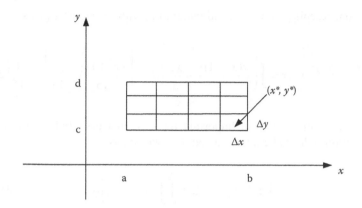

Figure 10.13 Rectangular region into smaller pieces.

This is just two iterated integrals to be calculated separately. The following examples show how to calculate these types of double integrals.

Example 10.12

Evaluate $\displaystyle\int_0^3 \int_0^1 (2xy - 4y)\, dy\, dx$

Solution: Do the inner integral with respect to y by fixing x, then do the x integral.

$$\int_0^3 \left[xy^2 - 2y^2 \right]_0^1 dx = \int_0^3 (x - 2)\, dx = \left[\frac{x^2}{2} - 2x \right]_0^3 = \frac{9}{2} - 6 = -\frac{3}{2}$$

Example 10.13

Evaluate $\displaystyle\int_0^3 \int_0^{\frac{\pi}{2}} x^2 \cos y\, dy\, dx$.

Solution:

$$\int_0^3 \left[x^2 \sin y \right]_0^{\frac{\pi}{2}} dx = \int_0^3 x^2\, dx = \left[\frac{x^3}{3} \right]_0^3 = 9$$

Example 10.14

Evaluate $\displaystyle\int_0^1 \int_0^1 xy\sqrt{x^2 + y^2}\, dy\, dx$ (more tricky problem).

Solution: Keeping x constant and integrating with respect to y gives

$$\int_0^1 \left[\frac{x\,(x^2 + y^2)^{\frac{3}{2}}}{3} \right]_0^1 dx = \int_0^1 \left(\frac{x(x^2+1)^{\frac{3}{2}}}{3} - \frac{x^4}{3} \right) dx = \left[\frac{(x^2+1)^{\frac{5}{2}}}{15} - \frac{x^5}{15} \right]_0^1 = \frac{1}{15}\left(2^{\frac{5}{2}} - 2 \right)$$

Fubini's theorem is an important theorem that essentially allows the integration order to be interchanged. It is written as

$$\int_a^b \int_c^d f(x,y)\,dy\,dx \equiv \int_c^d \int_a^b f(x,y)\,dx\,dy \qquad (10.13)$$

This is just saying that it does not matter if you sum x first or y first.

10.2.1.1   An Application of Double Integration

Find the volume under a surface $z = f(x,y)$ as shown in Figure 10.14. To compute this volume chop the region D into pieces. The volume of that "tower" piece is approximately:

$$\text{Vol} \approx \lim_{\Delta A \to 0} \sum_{\text{all pieces}} f(x^*, y^*)\Delta A$$

This approaches the volume below the surface:
Therefore,

$$\text{Volume} = \iint_D f(x,y)\,dA \qquad (10.14)$$

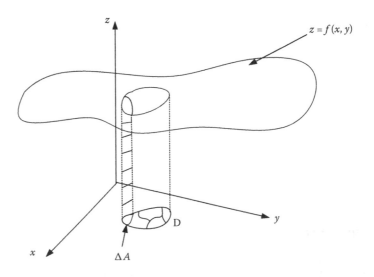

Figure 10.14 Volume under a surface.

Example 10.15

Find the volume of the solid in the first octant (see Figure 10.15) bounded by $z = 9 - x^2$ and $y = 4$.

Solution: What surface does $z = 9 - x^2$ look like? Since y is not present it can take on any value. This region is shown in Figure 10.15.

$$\text{Volume} = \iint_D f(x,y)\,dA = \iint_D (9 - x^2)\,dA$$

$$= \int_0^3 \int_0^4 (9 - x^2)\,dy\,dx$$

Integrating with respect to y and keeping x constant gives

$$\int_0^3 \left[9y - yx^2\right]_0^4 dx = \int_0^3 (36 - 4x^2)\,dx = \left[36x - \frac{4x^3}{3}\right]_0^3 = 72$$

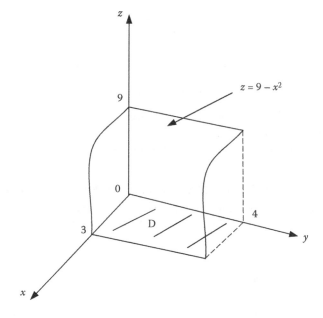

Figure 10.15 Volume bounded by the surfaces.

10.2.2 Double Integration Using Polar Coordinates

First. let's review concepts of polar coordinates. Figure 10.16 shows a point on the x-y plane. The point P can be identified using the usual (x, y) coordinates but also can be given by the angle θ with the x-axis and a distance r from the origin. Using this (r, θ) gives the *polar form* of a point in 2-D space.

Using basic trigonometry gives

$$x = r\cos\theta$$

$$y = r\sin\theta$$

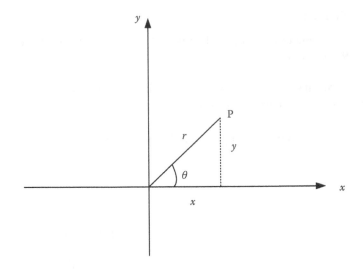

Figure 10.16 General point P in 2-D space.

Using the Pythagorean theorem gives

$$r = \sqrt{x^2 + y^2}$$

Note: Sometimes $\theta = tan^{-1}\left(\dfrac{y}{x}\right)$, but in general must be careful if the point is in a different quadrant, hence the need to use trigonometry to find the angle θ according to the problem being solved.

10.2.2.1   Using Polar Coordinates to Calculate Double Integrals

Suppose the region lies as shown in Figure 10.17. Calculate the area of the region D such that $\alpha \le \theta \le \beta$ and $r_1(\theta) \le r \le r_2(\theta)$. Again using the formula for general area gives

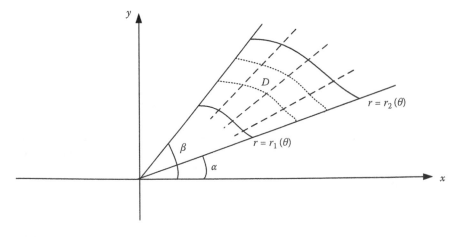

Figure 10.17 Area bounded by polar coordinates.

$$\iint_D f(x, y)\, dA \triangleq \lim_{\Delta A \to 0} \sum_{all\ pieces} f(x^*, y^*)\Delta A$$

$$= \lim_{\Delta r \to 0} \lim_{\Delta \theta \to 0} \sum_i \sum_j f\left(r_i^* \cos\theta_j^*,\ r_i^* \sin\theta_j^*\right) r\Delta r \Delta\theta$$

The r in $r\Delta r\Delta\theta$ is needed here for the change of variable (see later section defining Jacobian).

$$\text{Area} = \int_\alpha^\beta \int_{r_1}^{r_2} f(r\cos\theta,\ r\sin\theta) r\, dr\, d\theta \tag{10.15}$$

Example 10.16

Consider the disc $x^2 + y^2 \leq 4$ has charge density $\sigma(x,y) = 3x + x^2 + y^2$. Compute the total charge on the disc. The diagram of this region is shown in Figure 10.18.

Solution: The area is given by formula

$$A = \iint_D \left(3x + x^2 + y^2\right) dA$$

You could use traditional Cartesian coordinates x and y, but it is easier here to use polar coordinates r and θ.

Changing to polar coordinates, $x = r \cos\theta$, $y = r \sin\theta$, and $r^2 = x^2 + y^2$ with $dA = r\, dr\, d\theta$.

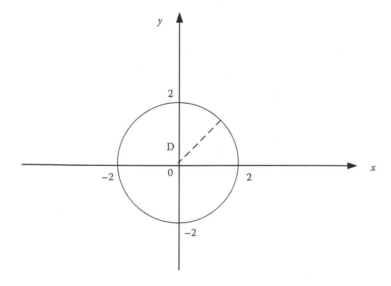

Figure 10.18 Disc with density $\sigma(x,y)$ and radius 2.

Also, the limits of integration now change to $0 \le \theta \le 2\pi$ and $0 \le r \le 2$.

$$\int_0^{2\pi} \int_0^2 (3r\cos\theta + r^2) r \, dr \, d\theta$$

Multiplying by r and integrating while keeping θ constant gives

$$= \int_0^{2\pi} \left[r^3 \cos\theta + \frac{1}{4} r^4 \right]_0^2 d\theta$$

$$= \int_0^{2\pi} \left[8\cos\theta + 4 \right] d\theta = 8\pi$$

Since the integral of both the $\cos\theta$ and $\sin\theta$ over a period are both equal to zero.

Example 10.17

Find the volume of the "snow cone" bounded by $z = \sqrt{x^2 + y^2}$ and $x^2 + y^2 + z^2 = 4$. Figure 10.19 shows the diagram of the region bounded.

Solution:

$$\text{Volume} = \iint \text{Upper surface} - \iint \text{Lower surface}$$

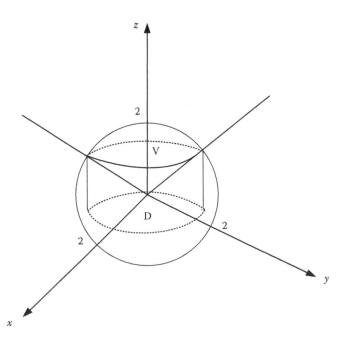

Figure 10.19 The snow cone bounded region.

Upper surface is the sphere $z = \sqrt{4 - x^2 - y^2}$.

Lower surface is the cone $z = \sqrt{x^2 + y^2}$.

Therefore,

$$\text{Volume} = \iint_D \left(\sqrt{4 - x^2 - y^2} - \sqrt{x^2 + y^2} \right) dA \qquad (10.16)$$

The region D is a circle, which is the intersection of the sphere with the cone as shown in Figure 10.20.

$$x^2 + y^2 + z^2 = 4 \quad \text{with} \quad z = \sqrt{x^2 + y^2}$$

This gives

$$x^2 + y^2 = 2 \quad \text{or} \quad x^2 + y^2 = \left(\sqrt{2} \right)^2$$

In polar coordinates Equation 10.16 becomes

$$\int_0^{2\pi} \int_0^{\sqrt{2}} \left(\sqrt{4 - r^2} - r \right) r \, dr \, d\theta$$

$$\int_0^{2\pi} \int_0^{\sqrt{2}} \left(r\sqrt{4 - r^2} - r^2 \right) dr \, d\theta$$

Which is just a standard integral with respect to r first then with respect to θ giving the volume as

$$\int_0^{2\pi} \left[-\frac{1}{3}(4 - r^2)^{\frac{3}{2}} - \frac{1}{3}r^3 \right]_0^{\sqrt{2}} d\theta = \int_0^{2\pi} \left(\frac{8 - 2^{\frac{5}{2}}}{3} \right) d\theta$$

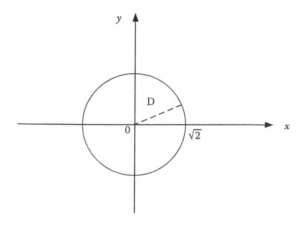

Figure 10.20 The circle that is the intersection of the sphere and cone.

$$\text{Volume} = \frac{2\pi}{3}\left(8 - 2^{\frac{5}{2}}\right)$$

10.2.3   General Regions

Previously, calculations of double integrals over a rectangular region were considered as shown in Figure 10.21.

The area of the region D is given by

$$\iint_D f(x,y)\,dA = \int_a^b \int_c^d f(x,y)\,dy\,dx = \int_c^d \int_a^b f(x,y)\,dx\,dy$$

Suppose the region D has the general form as shown in Figure 10.22. The area of the region D can be found as

$$\iint_D f(x,y)\,dA = \int_a^b \int_{g_1(x)}^{g_2(x)} f(x,y)\,dy\,dx$$

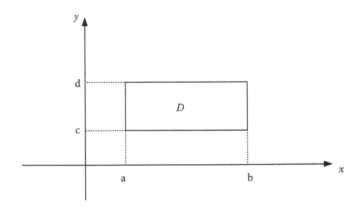

Figure 10.21 Rectangular region as area.

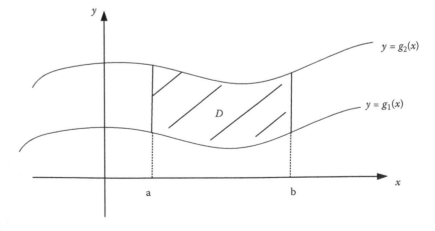

Figure 10.22 General region with area D.

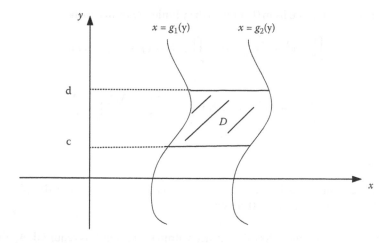

Figure 10.23 General region with area D.

Again here integrate out y first then x.

Also, if D has the following form as in Figure 10.23.
Again this is given by

$$\iint_D f(x,y)\,dA = \int_c^d \int_{g_1(y)}^{g_2(y)} f(x,y)\,dy\,dx$$

Here integrate out x first then y to find the required area.

Example 10.18

Integrate $f(x,y) = 4xy - x^2$ over the region D given in Figure 10.24.

Solution: Using the definition gives

$$\iint_D f(x,y)\,dA = \int_0^1 \int_0^{1-x} (4xy - x^2)\,dy\,dx$$

Note: Usually, the limits are the most difficult part of the integral to sort out correctly.

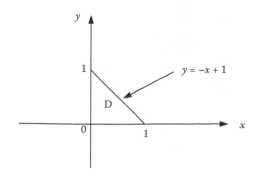

Figure 10.24 Triangular region D.

Here if x limits are from 0 to 1 then the y limits are from 0 to the line $y = 1 - x$.

$$= \int_0^1 \left[2xy^2 - x^2y \right]_0^{1-x} dx = \int_0^1 \left[2x(1-x)^2 - x^2(1-x) \right] dx$$

$$= \int_0^1 (2x - 5x^2 + 3x^3) dx = \left[x^2 - \frac{5x^3}{3} + \frac{3x^4}{4} \right]_0^1 = \frac{1}{12}$$

Example 10.19

Find the volume enclosed by the paraboloid $z = x^2 + 3y^2$ and the planes $x = 0$, $y = 1$, $y = x$, and $z = 0$ shown in Figure 10.25.

Solution: The volume below the surface within the region D is required. Again if the limits for x are from 0 to 1, then the limits for y are from 0 to $y = x$.

$$\iint_D (x^2 + 3y^2) \, dA = \int_0^1 \int_x^1 (x^2 + 3y^2) \, dy \, dx$$

$$= \int_0^1 \left[x^2 y + y^3 \right]_x^1 \, dx = \int_0^1 \left[(x^2 + 1) - 2x^3 \right] dx$$

$$= \left[\frac{x^3}{3} + x - \frac{x^4}{2} \right]_0^1 = \frac{5}{6}$$

The next section looks at triple integrals and some of their applications.

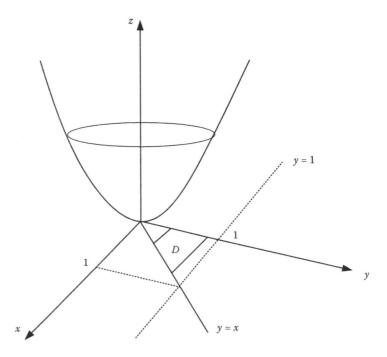

Figure 10.25 The volume enclosed under the paraboloid.

10.2.4 Triple Integrals

Consider a 3-D region E in space as shown in Figure 10.26.

What is meant by triple integration? This is exactly the same concept as double integration, that is, splitting the region into smaller volumes and summing.

$$\iiint_E f(x,y,z)\, dV \triangleq \lim_{\Delta V \to 0} \sum_{all\ pieces} f(x^*, y^*, z^*)\Delta V \qquad (10.17)$$

Again, there are similar applications for triple integrals as double integrals.

$$\mathrm{Vol}\,(E) = \iiint_E 1.\, dV \quad \text{with} \quad f(x^*, y^*, z^*) = 1$$

$$\mathrm{Mass}\,(E) = \iiint_E d(x,y,z)\, dV$$

In 2-D space the double integrals were calculated using either Cartesian or polar coordinates. These concepts can be extended to 3-D space.

To calculate triple integrals in 3-D space, different coordinate systems can be used, including Cartesian, cylindrical, or spherical coordinates depending on which will make the integration easier to compute.

The next section starts with evaluating triple integrals in Cartesian coordinates.

10.2.4.1 Cartesian Coordinates

Express the triple integrals as

$$\iiint_E f(x,y,z)\, dV = \iiint_{Limits} f(x,y,z)\, dz\, dy\, dx = \iiint_{Limits} f(x,y,z)\, dy\, dx\, dz$$

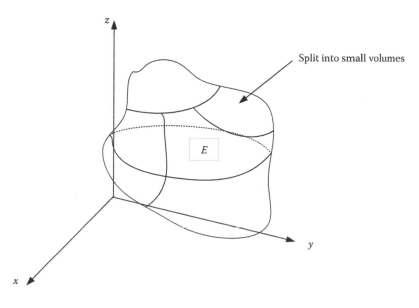

Figure 10.26 3-D volume enclosed by a region.

This integral can be done six different ways since the order of the integration can be taken with respect to the different variables x, y, or z as required.

Example 10.20

Let E be the region shown in Figure 10.27 which is bounded by the planes $x + y + z = 1$, $x = 0$, $y = 0$ and $z = 0$. Suppose E has mass density $d(x,y,z) = x^2 y$. Compute the mass of E of the region bounded by the above planes.

Solution: See Figure 10.27.
   Using the formula for the mass and replacing density $d(x,y,z) = x^2 y$ gives

$$\text{Mass } (E) = \iiint_E x.\ dV \ = \ \overset{1}{\underset{0}{\int}} \overset{1-x}{\underset{0}{\int}} \overset{1-x-y}{\underset{0}{\int}} x^2 y \, dz \, dy \, dx$$

Line in x-y plane   Plane $z = 1 - x - y$

Integrate out with respect to z first gives

$$= \int_0^1 \int_0^{1-x} x^2 y [z]_0^{1-x-y} \, dy \, dx = \int_0^1 \int_0^{1-x} x^2 y (1 - x - y) \, dy \, dx$$

Integrate out with respect to y gives

$$= \int_0^1 \left[\frac{x^2 y^2}{2} - \frac{x^3 y^2}{2} - \frac{x^2 y^3}{3} \right]_0^{1-x} dx$$

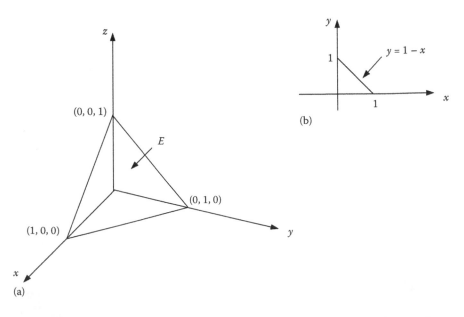

(a)

(b)

Figure 10.27 (a) The bounded volume E. (b) The triangular region in the x-y plane.

$$= \int_0^1 \left[\frac{1}{2} x^2 (1-x)^2 - \frac{1}{2} x^3 (1-x)^2 - \frac{1}{3} x^2 (1-x)^3 \right] dx$$

$$= \int_0^1 \left[\frac{1}{6} x^2 - \frac{1}{2} x^3 + \frac{1}{2} x^4 - \frac{1}{6} x^5 \right] dx$$

This is worked out using single-variable integration with respect to x and the answer turns out to be $\frac{1}{360}$.

10.2.5   3-D Coordinate Systems

In 3-D space the following coordinate systems can be used: Cartesian, cylindrical, and spherical coordinates. First considering cylindrical coordinates (θ, **r**, **z**) as shown in Figure 10.28. This is basically using polar coordinates plus the z coordinate. Relationships are given by

$$x = r \cos \theta$$

$$y = r \sin \theta$$

$$r = \sqrt{x^2 + y^2}$$

$$z = z$$

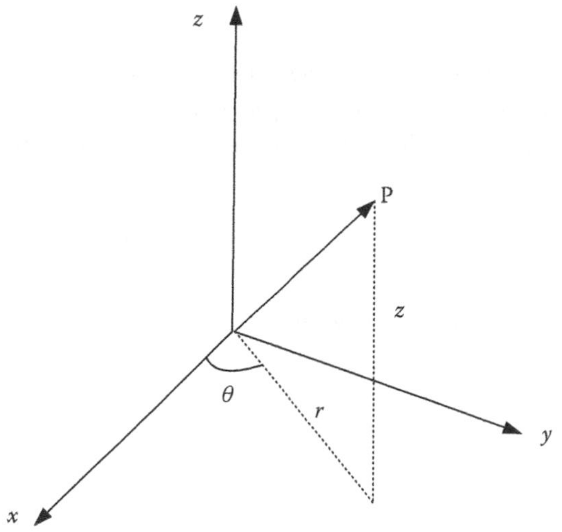

Figure 10.28 General point P in space in terms of cylindrical coordinates.

Now consider spherical coordinates (θ, φ, ρ). Here there are two angles and a distance to travel from the origin as shown in Figure 10.29. Relationships are given by the following

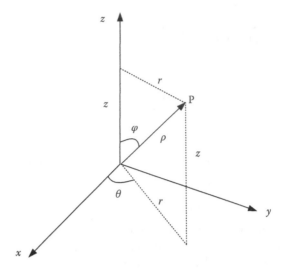

Figure 10.29 General point P in space in terms of spherical coordinates.

$$z = \rho \cos \varphi$$

$$r = \rho \sin \varphi$$

$$x = \rho \sin \varphi \cos \theta$$

$$y = \rho \sin \varphi \sin \theta$$

$$\rho = \sqrt{x^2 + y^2 + z^2}$$

10.2.5.1   Integrals in the New Coordinate Systems

As with the 2-D case when Cartesian coordinates were changed to polar coordinates, there was an introduction of the extra term known as the Jacobian and a similar term appears again in the 3-D change of coordinates as shown next.

Cylindrical:

Jacobian

$$\iiint_E f(x, y, z)\, dV = \iiint_{Limits!} f(r \cos \theta,\, r \sin \theta,\, z)\, r\, dz\, dr\, d\theta$$

Spherical:

Jacobian

$$\iiint_E f(x, y, z)\, dV = \iiint_{Limits!} f(\rho \sin \varphi \cos \theta,\, \rho \sin \varphi \sin \theta,\, \rho \cos \varphi)\, \rho^2 \sin \varphi\, d\rho\, d\varphi\, d\theta$$

These formulae can be used to find and prove the volumes of standard shapes.

Example 10.21

Find the volume of a sphere (E) of radius r and center the origin as shown in Figure 10.30.

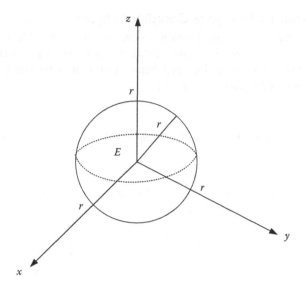

Figure 10.30 Sphere of radius r and center origin.

Solution: Volume is given by the formula

$$\text{Vol }(E) = \iiint_E 1.\,dv$$

Using spherical coordinates gives

$$= \int_0^{2\pi}\int_0^{\pi}\int_0^{r} 1\rho^2 \sin\varphi\, d\rho\, d\varphi\, d\theta$$

Integrate out w.r.t $d\rho$ gives

$$= \int_0^{2\pi}\int_0^{\pi}\left[\frac{\rho^3}{3}\sin\varphi\right]_0^{r} d\varphi\, d\theta = \int_0^{2\pi}\int_0^{\pi}\frac{r^3}{3}\sin\varphi\, d\varphi\, d\theta$$

Integrate out w.r.t $d\varphi$ gives

$$= \int_0^{2\pi}\left[-\frac{r^3}{3}\cos\varphi\right]_0^{\pi} d\theta = \int_0^{2\pi}\frac{2r^3}{3}\, d\theta$$

Integrate out w.r.t $d\theta$ gives

$$= \frac{2r^3}{3}\times 2\pi = \frac{4}{3}\pi r^3$$

So,

$$\text{Vol}(E) = \frac{4}{3}\pi r^3$$

This is just the usual expression for the volume of a sphere of radius r.

10.2.6   General Change of Coordinate Systems

It can be important in some applications in engineering to be able to change to a different coordinate system to enable the integration to be easily carried out. The next section shows how this is done in general, starting with the one-variable case.

For the one-variable case, let $x = g(u)$

$$\int_a^b f(x)\,dx = \int_{g^{-1}(a)}^{g^{-1}(b)} f\big[g(u)\big]g'(u)\,du = \int_{lower\ limit}^{upper\ limit} f\big[g(u)\big]\big|g'(u)\big|\,du$$

For the two-variable case, making a change of variables to u and v gives the transformation shown in Figure 10.31:

$$\iint_D f(x, y)\,dy\,dx = \iint_{T(D)} f[x(u, v), y(u, v)]\left|\frac{\partial(x, y)}{\partial(u, v)}\right|\,dv\,du$$

Jacobian

where

$$\frac{\partial(x, y)}{\partial(u, v)} = \det\begin{pmatrix} \dfrac{\partial x}{\partial u} & \dfrac{\partial y}{\partial u} \\[2ex] \dfrac{\partial x}{\partial v} & \dfrac{\partial y}{\partial v} \end{pmatrix}$$

The idea behind this coordinate transformation is shown in Figure 10.32.

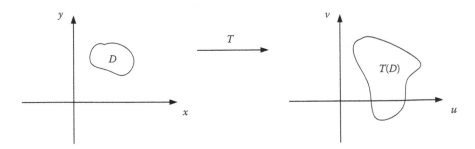

Figure 10.31 How a region D in the x-y space gets transformed into the u-v space.

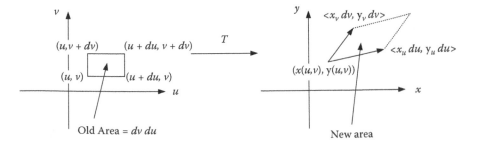

Figure 10.32 Area gets mapped to a new area under a general transformation.

Using the Taylor series expansion gives

$$\begin{cases} x(u+du,v) \approx x(u,v) + du\, x_u(u,v) \\ y(u+du,v) \approx y(u,v) + du\, y_u(u,v) \\ x(u,v+dv) \approx x(u,v) + dv\, x_v(u,v) \\ y(u,v+dv) \approx y(u,v) + dv\, y_v(u,v) \end{cases}$$

The new area is

$$= \left| \det \begin{bmatrix} x_u du & y_u du \\ x_v dv & y_v dv \end{bmatrix} \right|$$

$$= \left| \det \begin{bmatrix} x_u & y_u \\ x_v & y_v \end{bmatrix} \right| dv du$$

$$= \left| \frac{\partial(x,y)}{\partial(u,v)} \right| dv du$$

Example 10.22

Evaluate $\displaystyle\iint_D (64xy)\, dA$, where D is the parallelogram with vertices, $(-1,3)$, $(1,-3)$, $(3,-1)$, and $(1,5)$ as shown in Figure 10.33.

Solution: The region is complicated using just the x and y coordinates system. So changing variables can make things easier. Let

$$u = y - x$$

$$v = y + 3x$$

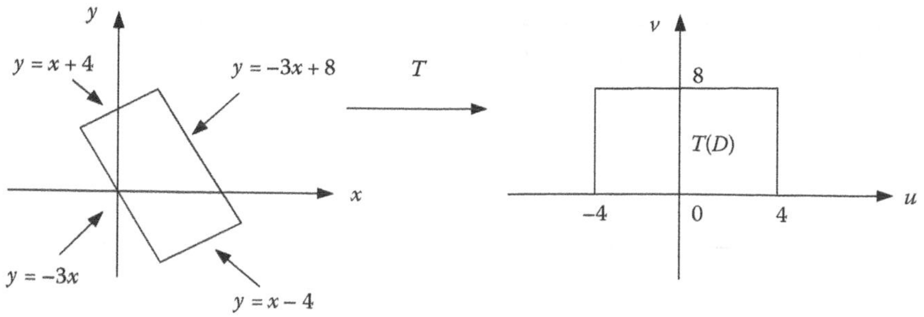

Figure 10.33 Area transformation with variable change.

These can be solved to find x and y and give

$$x = \frac{1}{4}(v - u) \quad \text{and} \quad y = \frac{1}{4}(3u + v)$$

So in this change of coordinate system of u and v gives a nice rectangular region in $u - v$ space.

The transformation mapping becomes

$$
\begin{array}{lll}
y = x - 4 & \text{maps to} & u = -4 \\
y = x + 4 & \text{maps to} & u = 4 \\
y = -3x + 8 & \text{maps to} & v = 8 \\
y = -3x & \text{maps to} & v = 0
\end{array}
$$

Now the change of variables gives

$$\iint_D (64xy)\,dA = \int_{-4}^{4} \int_{0}^{8} 64\left[\frac{1}{4}(v-u)\right]\left[\frac{1}{4}(3u+v)\right]\left|\frac{\partial(x,y)}{\partial(u,v)}\right|\,dv\,du$$

So

$$\frac{\partial(x,y)}{\partial(u,v)} = \det\begin{pmatrix} \dfrac{\partial x}{\partial u} & \dfrac{\partial x}{\partial v} \\[2mm] \dfrac{\partial y}{\partial u} & \dfrac{\partial y}{\partial v} \end{pmatrix} = \det\begin{pmatrix} -\dfrac{1}{4} & \dfrac{1}{4} \\[2mm] \dfrac{3}{4} & \dfrac{1}{4} \end{pmatrix} = -\frac{1}{4}$$

Now the integral becomes

$$\int_{-4}^{4}\int_{0}^{8} 4(v^2 + 2uv - 3u^2)\left|-\frac{1}{4}\right|\,dv\,du$$

$$= \int_{-4}^{4}\left[\frac{v^3}{3} + uv^2 - 3u^2v\right]_0^8 du = \int_{-4}^{4}\left(\frac{512}{3} + 64u - 24u^2\right)du = \frac{1024}{3}$$

Note: Sometimes it is easier to use the following to calculate

$$\frac{\partial(x,y)}{\partial(u,v)} = \frac{1}{\dfrac{\partial(u,v)}{\partial(x,y)}}$$

The above analysis can be further generalized to a three-variable case as follows:

$$\iiint_E f(x,y,z)\,dz\,dy\,dx = \iiint f\big(x(u,v,w)......\big)\left|\frac{\partial(x,y,z)}{\partial(u,v,w)}\right|dw\,dv\,du$$

with $x = x(u,v,w)$, $y = y(u,v,w)$, $z = z(u,v,w)$, where

$$\frac{\partial(x,y,z)}{\partial(u,v,w)} = \det\begin{pmatrix} x_u & x_v & x_w \\ y_u & y_v & y_w \\ z_u & z_v & z_w \end{pmatrix}.$$

So cylindrical and spherical coordinates are examples of three-variable problems and their Jacobians can be found using the aforementioned formula.

10.3 Applications

There are many physical applications of double and triple integration, most of which depend on the idea of splitting a region into smaller pieces and summing over all the pieces. Start by looking at applications of double integrals, which shows the principle involved.

10.3.1 Application of Double Integration

Remember what is meant by the double integration (see Figure 10.34). The definition of the double integral is

$$\iint_D f(x,y)\,dA \triangleq \lim_{\Delta A \to 0} \sum_{all\ pieces} f(x^*,y^*)\Delta A$$

The region is chopped into pieces. Then pick a point (x^*, y^*) and sum over all the pieces with $f(x^*, y^*)$ multiplied by the width of all the pieces ΔA.

This forms the basis of all applications as shown in the next few examples.

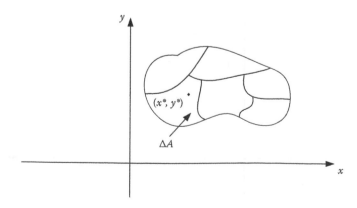

Figure 10.34 General region in the x-y plane.

Example 10.23: Volume under a Surface

Find the volume under a surface $z = f(x,y)$ shown in Figure 10.35.

Solution: The approximate volume under the single column is $f(x^*,y^*)\Delta A$. Then the total volume is given by

$$\text{Volume} \approx \lim_{\Delta A \to 0} \sum_{all\ pieces} f(x^*, y^*)\Delta A$$

$$\text{Volume} = \iint_D f(x, y)\, dA \qquad (10.18)$$

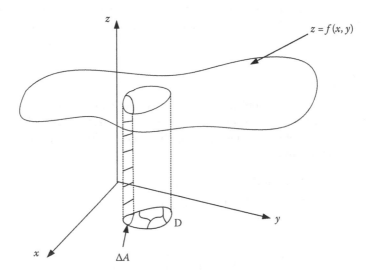

Figure 10.35 Volume under a surface.

Example 10.24: Area of a Bounded Region

Find the area of a region D shown in Figure 10.36.

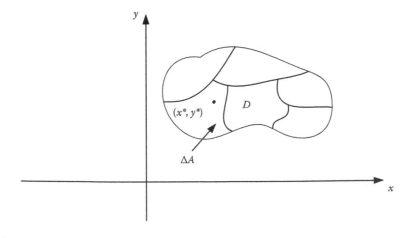

Figure 10.36 Area of a region.

Solution: Here, if $f(x^*, y^*) = 1$, then one has just the area of the small piece and then summing gives

$$\text{Area} \approx \lim_{\Delta A \to 0} \sum_{all\ pieces} 1.\Delta A$$

$$\text{Area} = \iint\limits_D 1\, dA \qquad\qquad (10.19)$$

Example 10.25: Mass of a Region

Find the mass of a region D shown in Figure 10.37, given that the mass density is $d(x,y)$ (mass/area).

Solution:

$$\text{Mass} \approx \lim_{\Delta A \to 0} \sum_{all\ pieces} d(x^*, y^*)\Delta A$$

$$\text{Mass} = \iint\limits_D d(x, y)\, dA \qquad\qquad (10.20)$$

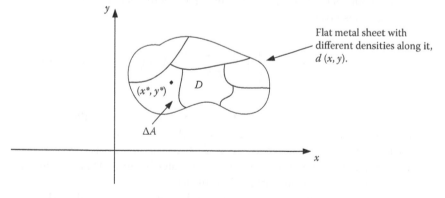

Figure 10.37 Mass of a region.

Example 10.26: Radiative Heat Transfer between Surfaces Using View Factors

Calculating the radiative heat transfer between different surfaces is an important area in many fields of engineering. In fire engineering, it can be an important consideration when calculating the radiative heat transfer from one building surface to nearby buildings to see if nearby buildings are in danger of catching fire.

Consider two finite surface areas A_i and A_j with small infinitesimal areas dA_i and dA_j with normal vectors n_i and n_j, respectively, as shown in Figure 10.38. The view factor between two finite areas A_i and A_j is denoted

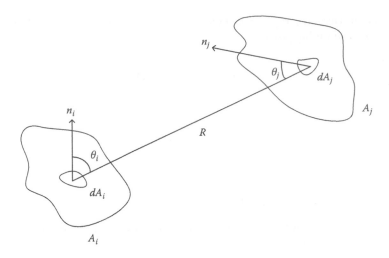

Figure 10.38 View factors between two finite surfaces.

by F_{ij} and defined as the fraction of the radiation leaving the surface i that is intercepted then by the surface j. This can be represented as

$$F_{ij} = \frac{[\textit{Intensity leaving i and hitting j}]}{\left[\textit{Total intensity leaving i}\right]}$$

The mathematical formula for this view factor needs some work to derive it and is given as

$$F_{ij} = \frac{1}{A_i} \int\limits_{A_i} \int\limits_{A_j} \frac{\cos\theta_i \cos\theta_j}{\pi R^2} dA_j \, dA_i \tag{10.21}$$

Essentially, this double integral arises by taking a particular dA_i and calculating its contributions to each of the dA_j and then summing all these contributions, which is the inner integral. Then the process is repeated for all the dA_i and summing, which is then the outer integral. This then gives the total contribution from surface i to surface j.

From a practical basis, this can involve lots of computational work and some simplifications can be made using view factor algebra to avoid using Equation 10.21. Simplifications using properties of reciprocity and summation can help in calculating view factors for multisurfaces.

10.3.2 Application of Triple Integration (Center of Mass)

The region shown in Figure 10.39 has a mass density of $d(x,y,z)$. Then the mass of the region is given by

$$\text{Mass}(E) = M = \iiint\limits_{E} d(x,y,z) dV \tag{10.22}$$

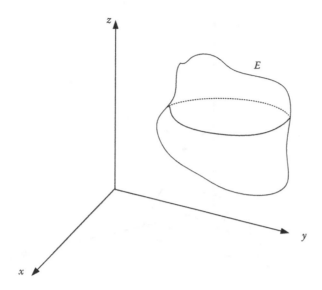

Figure 10.39 Center of mass of a region.

Therefore, the center of mass has position $(\overline{x}, \overline{y}, \overline{z})$, where

$$\overline{x} = \frac{1}{M} \iiint_E x\, d(x, y, z)\, dV \qquad (10.23)$$

$$\overline{y} = \frac{1}{M} \iiint_E y\, d(x, y, z)\, dV \qquad (10.24)$$

$$\overline{z} = \frac{1}{M} \iiint_E z\, d(x, y, z)\, dV \qquad (10.25)$$

Example 10.27: Center of mass of a body

Compute the center of mass of the region E shown in Figure 10.40 enclosed by $z = 1 - x^2 - y^2$ and $z = 0$. Assume a constant mass density $d = 1$.

Solution: See Figure 10.40. By radial symmetry, the center of mass must be on the z axis with more mass lower down so it should be closer to the $z = 0$ plane. By symmetry, $\overline{x} = 0$ and $\overline{y} = 0$ are already known. Now to work out $\overline{z}$ use the formula given by Equation 10.25 with $d = 1$ gives

$$\overline{z} = \frac{1}{M} \iiint_E z\, d(x, y, z)\, dV = \frac{1}{M} \iiint_E z\, d\, dV$$

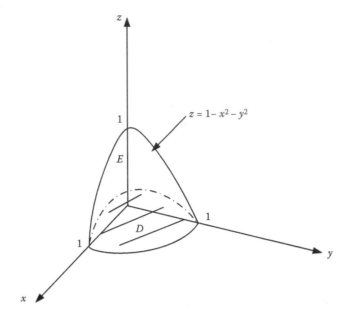

Figure 10.40 Region enclosed by the surfaces.

First, work out the mass M of the region using the mass formula given by Equation 10.22:

$$M = \iiint_E 1\, dV$$

$$= \iint_D \int_0^{1-x^2-y^2} 1\, dz\, dy\, dx$$

$$= \iint_D (1 - x^2 - y^2)\, dy\, dx$$

Now D is the circle of radius 1 in the x-y plane. So changing to polar coordinates gives

$$\int_0^{2\pi} \int_0^1 (1 - r^2)\, r\, dr\, d\theta = \frac{\pi}{2}$$

This is the expression for mass M.

For the center of mass calculation, use Equation 10.25 to give

$$\bar{z} = \frac{1}{M} \iiint_E z\, dV$$

$$= \frac{2}{\pi} \iint_D \int_0^{1-x^2-y^2} z\, dz\, dy\, dx$$

$$= \frac{2}{\pi} \iint_D \frac{1}{2}(1-x^2-y^2)^2\, dy\, dx$$

Changing to polar coordinates with $r^2 = x^2 + y^2$ gives

$$= \frac{1}{\pi} \int_0^{2\pi} \int_0^1 (1-r^2)^2 r\, dr\, d\theta = \frac{1}{\pi} \int_0^{2\pi} \left[-\frac{1}{6}(1-r^2)^3 \right]_0^1 d\theta$$

$$= \frac{1}{\pi} \int_0^{2\pi} \frac{1}{6} d\theta = \frac{1}{3}$$

$\therefore$ center of mass is at $(\bar{x}, \bar{y}, \bar{z}) = \left(0, 0, \frac{1}{3} \right)$.

Problems

10.1 Given that $f(x,y) = x^3y + y^3$, evaluate the following.

a. $\left. \dfrac{\partial f}{\partial x} \right|_{(1,1)}$ b. $\left. \dfrac{\partial f}{\partial y} \right|_{(2,1)}$

10.2 Given that $f(x,y) = x\sin(xy)$, evaluate the following.

a. f_x b. f_y c. f_{xy} d. f_{yx}

10.3 Given that $f(x,y) = xy^2 - 5xy$, calculate the directional derivative of f in the direction of $\langle 3,1 \rangle$ at the point $(1,1)$.

10.4 Evaluate the following double integral:

$$\int_0^3 \int_0^1 (x^2y - 5x)\, dy\, dx$$

10.5 Find the volume of the solid tetrahedron enclosed by the plane $2x + y + z = 4$ and the coordinate planes.

10.6 In Figure 10.41, a cylinder of height H and radius R is shown. Show that the volume of the cylinder is given by the formula $V = \pi R^2 H$.

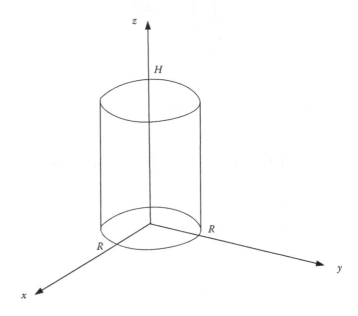

Figure 10.41 A general cylinder of height H and radius R.

11 Vector Calculus

11.1 Differentiation and Integration of Vectors

Vector calculus deals with the differentiation and integration of vector fields in two- and three-dimensional space. In this chapter important concepts such as the different types of line integrals are first considered and then how closed line integrals can be equivalent to double integrals over a region known as Green's theorem. Further important ideas on the gradient and curl of vector fields are developed leading to surface integrals and their applications to describe fluid flow and many different force fields that occur naturally in the world.

11.1.1 Derivatives of Vector Functions

A curve is defined in 3-D space in parametric form as $\overline{r}(t)$, where

$$\overline{r}(t) = \langle f(t), g(t), h(t) \rangle$$

Note: ⟨u,v,w⟩ represents a general three-dimensional vector.

The derivative of $r(t)$ with respect to time t is defined as $\overline{r}'(t)$ and is given by

$$\overline{r}'(t) = \lim_{\Delta t \to 0} \frac{\overline{r}(t + \Delta t) - \overline{r}(t)}{\Delta t} \tag{11.1}$$

This is a tangent vector to a space curve. Also, this can be thought of as the instantaneous velocity, as shown in Figure 11.1.

As, $\Delta t \to 0$ the vector $\overline{PQ}$ tends to the tangent vector at the point P.

$$= \lim_{\Delta t \to 0} \left[\frac{\left[f(t + \Delta t)\underline{i} + g(t + \Delta t)\underline{j} + h(t + \Delta t)\underline{k} \right] - \left[f(t)\underline{i} + g(t)\underline{j} + h(t)\underline{k} \right]}{\Delta t} \right]$$

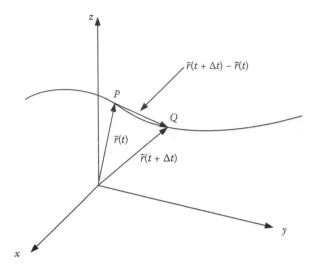

Figure 11.1 The tangent vector to a curve.

$$= \lim_{\Delta t \to 0} \left[\frac{\left[\left[f(t+\Delta t) - f(t) \right] \underline{i} + \left[g(t+\Delta t) - g(t) \right] \underline{j} + \left[h(t+\Delta t) - h(t) \right] \underline{k} \right]}{\Delta t} \right]$$

$$= \lim_{\Delta t \to 0} \left[\frac{\left[f(t+\Delta t) - f(t) \right] \underline{i}}{\Delta t} + \frac{\left[g(t+\Delta t) - g(t) \right] \underline{j}}{\Delta t} + \frac{\left[h(t+\Delta t) - h(t) \right] \underline{k}}{\Delta t} \right]$$

So,

$$\bar{r}'(t) = f'(t)\underline{i} + g'(t)\underline{j} + h'(t)\underline{k} \qquad (11.2)$$

Therefore, to find the derivative of $r(t)$ with respect to time t, just differentiate each of the components of $r(t)$, that is, given that

$$\bar{r}(t) = \langle f(t),\ g(t),\ h(t) \rangle$$

then

$$\bar{r}'(t) = \langle f'(t),\ g'(t),\ h'(t) \rangle.$$

Example 11.1

If $\bar{r}(t) = \langle t, t^2, t^3 \rangle$, then

$$\bar{r}'(t) = \langle f'(t),\ g'(t),\ h'(t) \rangle = \langle 1,\ 2t,\ 3t^2 \rangle$$

This is the directional vector of the tangent line space curve at any point t for the vector function $\bar{r}(t)$.

Also differentiating a second time gives

$$\bar{r}''(t) = \langle f''(t),\ g''(t),\ h''(t) \rangle = \langle 0, 2, 6t \rangle$$

11.1.2   Integrating Vector Functions

Integrals or antiderivatives work like normal integrals do as shown in the next few examples.

Example 11.2

$$\int_0^1 \left(t\underline{i} + t^2 \underline{j} + t^3 \underline{k} \right) dt$$

$$= \left[\frac{1}{2} t^2 \underline{i} + \frac{1}{3} t^3 \underline{j} + \frac{1}{4} t^4 \underline{k} \right]_0^1$$

$$= \frac{1}{2} \underline{i} + \frac{1}{3} \underline{j} + \frac{1}{4} \underline{k} \quad \text{or} \quad \langle \frac{1}{2}, \frac{1}{3}, \frac{1}{4} \rangle$$

Example 11.3

$$\int \left(\frac{1}{t} \underline{i} + t^2 \underline{j} + e^{3t} \underline{k} \right) dt = \ln t \, \underline{i} + \frac{t^3}{3} \underline{j} + \frac{e^{3t}}{3} \underline{k} + \underline{a}$$

Where $\underline{a}$ is a constant vector, that is, $\left(\underline{a} = c_1 \underline{i} + c_2 \underline{j} + c_3 \underline{k} \right)$.

Example 11.4

Given that $\bar{r}(t) = 2\underline{i} + 4t\underline{j} - 6t^2 \underline{k}$ and $\bar{r}(0) = \underline{j} + \underline{k}$, find $\bar{r}(t)$.

Solution:

$$\bar{r}(t) = \int \left(2\underline{i} + 4t\underline{j} - 6t^2 \underline{k} \right) dt$$

$$\bar{r}(t) = 2t\underline{i} + 2t^2 \underline{j} - 2t^3 \underline{k} + \underline{c}$$

where $\underline{c}$ is a constant vector.

$$\bar{r}(0) = 0 + 0 + 0 + \underline{c} = \underline{j} + \underline{k}$$

Therefore,

$$\bar{r}(t) = 2t\underline{i} + 2t^2 \underline{j} - 2t^3 \underline{k} + \underline{j} + \underline{k} \quad \text{or} \quad \bar{r}(t) = 2t\underline{i} + (2t^2 + 1)\underline{j} + (1 - 2t^3)\underline{k}$$

11.2   Vector Fields

Previously, the functions had different input variables but only one output variable. For example,

$f(x)$ 1 input, 1 output

$f(x, y)$ 2 inputs, 1 output

$f(x, y, z)$ 3 inputs, 1 output

Now, looking at *vector fields*, these can be represented as follows:

$$\overline{F}(x,y) = \langle M(x,y), N(x,y) \rangle \quad \text{2 inputs, 2 outputs}$$

$$\overline{F}(x,y,z) = \langle M(x,y,z), N(x,y,z), P(x,y,z) \rangle \quad \text{3 inputs, 3 outputs}$$

a vector field is a more general function. So to every point in the *x-y* plane or 3-D space it assigns a *vector*.

Some simple vector fields can be seen by graphing them as follows.

Example 11.5

What does the vector field $\overline{F}(x, y) = \langle x, y \rangle$ look like? See Figure 11.2.

What is the magnitude of the vector field, that is, $\left| \overline{F}(x,y) \right|$? The magnitude of a vector is given by the square root of all the components squared and added together. Therefore,

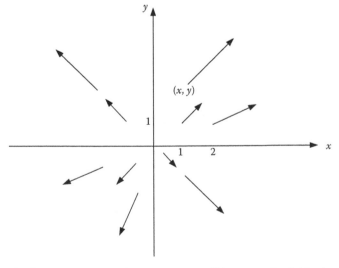

Look at points and see what vector is given. Take a general point (x, y) in space and this vector from the origin is the vector you put at the location (x, y).

Figure 11.2 The vector field for $\overline{F}(x, y) = \langle x, y \rangle$.

$$\left|\bar{F}(x,y)\right| = \sqrt{x^2 + y^2} = r$$

So the magnitude of the vector field at any position is just the distance from the origin to that point, pointing radially outward as shown in Figure 11.2.

Example 11.6

Sketch the vector field given by

$$\bar{F}(x,y) = \frac{\langle x, y \rangle}{\sqrt{x^2 + y^2}}$$

which is not defined at (0,0).

Solution: Again this is the same vector field as in Example 11.5 but now the magnitude is always equal to unity as shown in Figure 11.3 (the arrows are all of length 1 unit).

The magnitude of the vector field is given by

$$\left|\bar{F}(x,y)\right| = \frac{\left|\langle x, y \rangle\right|}{\sqrt{x^2 + y^2}} = \frac{\sqrt{x^2 + y^2}}{\sqrt{x^2 + y^2}} = 1$$

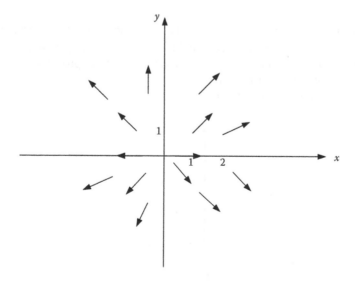

Figure 11.3 The vector field for $\bar{F}(x,y) = \dfrac{\langle x, y \rangle}{\sqrt{x^2 + y^2}}$.

Example 11.7

Sketch the vector field given by, $\bar{F}(x,y) = \langle y, 0 \rangle$.

Solution: This vector field shown in Figure 11.4 is known as *shear flow* in fluid dynamics and models flow of water near boundaries.

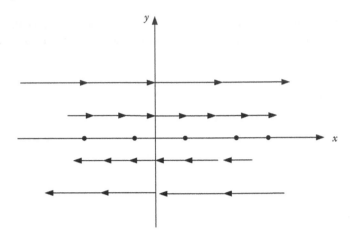

Figure 11.4 The vector field for $\bar{F}(x,y) = \langle y,0 \rangle$.

Example 11.8

Consider the vector force field for gravity. For gravitational attraction the magnitude of the force varies as reciprocal of the distance squared, that is,

$$\text{Force of gravity} = \left|\bar{F}(x,y,z)\right| \sim \frac{1}{r^2}(\text{distance squared})$$

It turns out that the vector field for the force due to gravity can be arrived at by starting with the basic vector field $\bar{F}(x,y,z) = \langle x,y,z \rangle$, multiplying by a constant and a negative sign, and making the modulus proportional to the reciprocal of the distance squared.

The sketch is shown in Figure 11.5.

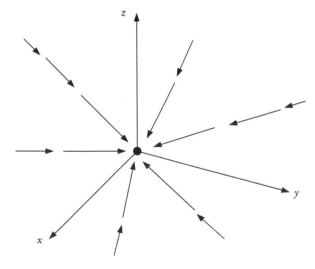

Figure 11.5 The vector field for gravitational attraction force.

The vector field representation is given as

$$\overline{F}(x,y) = \frac{-k\langle x,y,z\rangle}{(x^2+y^2+z^2)^{\frac{3}{2}}}$$

11.3 Line Integrals

The basic idea of a line integral is to integrate along a curve C in 2-D or 3-D, as shown in Figure 11.6.

11.3.1 The *ds*-Type Integral

$$\int_C f(x,y)\,ds = \lim_{\Delta s \to 0} \sum_{all\ pieces} f(x^*,y^*)\Delta s$$

where the Δs is the length of a small piece.

Applications of the *ds*-type integral include the following:

1. Length of the curve is given by

$$\text{Length} = \int_C 1.ds \qquad (11.3)$$

2. If $d(x, y)$ is the mass density (mass/length) of the wire, then the total mass of the wire is given by

$$\text{Mass of wire} = \int_C d(x,y)\,ds \qquad (11.4)$$

Note: These formulae can be used to calculate how long or heavy are the steel cables needed in the design of suspension bridges.

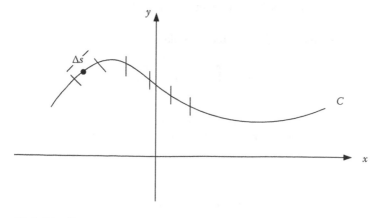

Figure 11.6 The line integral in 2-D space.

To compute the *ds*-type of integral, first, the curve C, as shown in Figure 11.7, needs to be parameterized.

Here, $x = x(t)$, $y = y(t)$, and $a \leq t \leq b$. Then *ds* can be found using the Pythagorean theorem as

$$ds = \sqrt{dx^2 + dy^2}$$

Then

$$\int_C f(x,y)\,ds = \int_C f(x,y)\sqrt{dx^2 + dy^2}$$

From which the integral in parametric form becomes

$$\int_C f(x,y)\,ds = \int_a^b f(x(t),y(t))\sqrt{\left(\frac{dx}{dt}\right)^2 + \left(\frac{dy}{dt}\right)^2}\,dt \qquad (11.5)$$

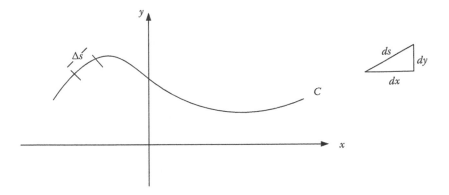

Figure 11.7 Parameterization of the curve.

Example 11.9

If the density of a wire is $d(x,y) = x$, compute its mass when the wire is the quarter circle of radius 2 as shown in Figure 11.8.

Solution: The mass is given by the formula

$$\text{Mass} = \int_C d(x,y)\,ds = \int_C x\,ds \qquad (11.6)$$

To parameterize the curve, since the wire is in the form of a circle, then the natural parameter is in terms of the angle made, that is, *t*.

$$x(t) = 2\cos t \qquad \dot{x}(t) = -2\sin t$$

$$y(t) = 2\sin t \qquad \dot{y}(t) = 2\cos t$$

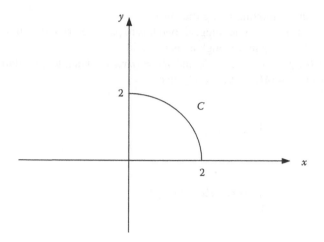

Figure 11.8 A wire in the form of a quarter circle.

$$0 \le t \le \frac{\pi}{2}$$

The mass is given by Equation 11.6 as

$$\text{Mass} = \int_{0}^{\frac{\pi}{2}} 2\cos t \sqrt{(-2\sin t)^2 + (2\cos t)^2} \, dt$$

$$= \int_{0}^{\frac{\pi}{2}} 4\cos t \, dt = 4[\sin t]_{0}^{\frac{\pi}{2}} = 4$$

11.3.2 The $d\bar{r}$ – Type Integral

Here a vector field acts in the region given by $\bar{F}(x,y)$. Again, the line is split into small segments as shown in Figure 11.9.

The component of the vector field in the direction of the line is given by $\bar{F}(x^*, y^*).\Delta\bar{r}$, so summing along the whole line gives

$$\int_{C} \bar{F}(x,y).d\bar{r} = \lim_{|\Delta\bar{r}| \to 0} \sum_{\text{all pieces}} \bar{F}(x^*, y^*).\Delta\bar{r} \qquad (11.7)$$

A useful application of this type of $d\bar{r}$ integral is the calculation of the *work done* along a curve through a vector field.

So, if $\bar{F}(x,y)$ is a force field, then

$$\text{Work} = \int_{C} \bar{F}.d\bar{r} \qquad (11.8)$$

is the work done in moving along the curve.

To compute this $d\bar{r}$ -type integral, one has to parameterize the curve C. Again Figure 11.10 shows a path through a vector field.

Let $x = x(t)$, $y = y(t)$, $a \leq t \leq b$, and $d\bar{r} = \langle dx, dy \rangle$. Then to calculate the work done with $\bar{F}(x, y) = \langle M(x, y), N(x, y) \rangle$ gives

$$= \int_C \bar{F}(x, y).d\bar{r}$$

$$= \int_C M(x, y)dx + N(x, y)dy$$

$$= \int_a^b \left[M(x(t), y(t)) \frac{dx}{dt} + N(x(t), y(t)) \frac{dy}{dt} \right] dt \tag{11.9}$$

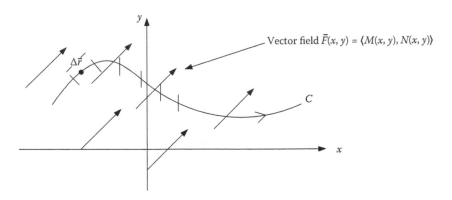

Figure 11.9 A path traveled through a vector field.

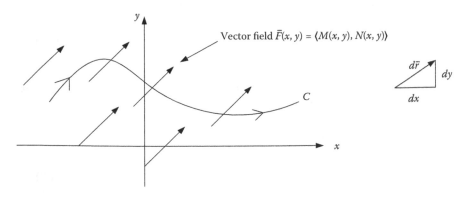

Figure 11.10 Parameterization of the path and vector field.

Example 11.10

If there is a force field $\bar{F}(x, y) = \langle x^2, y \rangle$, a particle travels along a curve C (i.e., $y = -x^2 + 1$) as shown in Figure 11.11.

So, the particle moves along the curve C to find the work done using Equation 11.8 gives

$$\text{Work} = \int_C \bar{F}.d\bar{r}$$

$$= \int_C x^2\, dx + y\, dy$$

Now to parameterize the curve C, one way is to let

$$x = t \qquad\qquad \frac{dx}{dt} = 1$$

$$y = -t^2 + 1 \qquad\qquad \frac{dy}{dt} = -2t$$

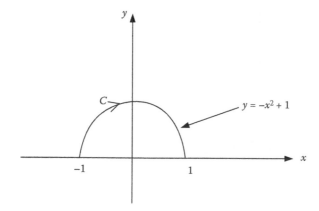

Fgure 11.11 Path traveled by the particle.

And then let t be between $-1 \le t \le 1$.
Using Equation 11.9 gives

$$\int_C \bar{F}(x, y).d\bar{r} = \int_a^b \left[M(x(t), y(t))\frac{dx}{dt} + N(x(t), y(t))\frac{dy}{dt} \right] dt$$

$$= \int_{-1}^{1} \left[(t)^2(1) + (-t^2 + 1)(-2t) \right] dt$$

$$= \int_{-1}^{1} \left[2t^3 + t^2 - 2t \right] dt = \frac{2}{3}$$

11.3.3   Summary of Results

A general path in 3-D space is shown in Figure 11.12.

The two types of line integrals can be summarized by the formulae as follows:

ds–type:

$$\int_C f(x,y,z)\,ds = \int_a^b f(x(t),y(t),z(t))\sqrt{\left(\frac{dx}{dt}\right)^2 + \left(\frac{dy}{dt}\right)^2 + \left(\frac{dz}{dt}\right)^2}\,dt$$

$d\bar{r}$ – type:

$$\int_C \bar{F}(x,y,z).d\bar{r} = \int_a^b \left[M(x(t),y(t),z(t))\frac{dx}{dt} + N(x(t),y(t),z(t))\frac{dy}{dt} + P(x(t),y(t),z(t))\frac{dz}{dt} \right]dt$$

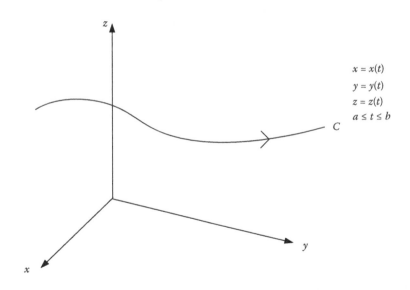

Figure 11.12 A general path in space.

11.4   Gradient Fieds

A special type of vector field is known as a *gradient field*.

Given $f(x,y)$, defining a vector field, the $\bar{F}(x,y) = \nabla f$

∇f is called its "gradient field."

Example 11.11

1. Given the function $f(x,y) = xy + y^3$, the gradient field is

$$\bar{F}(x,y) = \nabla f = \langle f_x, f_y \rangle = \langle y, x + 3y^2 \rangle$$

2. Given the function $f(x,y) = \frac{1}{2}(y-x^2)$, then the gradient field is given by

$$\bar{F}(x,y) = \nabla f = \langle f_x, f_y \rangle = \langle -x, \frac{1}{2} \rangle$$

Now $f(x,y) = \frac{1}{2}(y-x^2)$. The level curves for this function, that is,

$$f = 0 = \frac{1}{2}(y-x^2) \qquad y = x^2$$

$$f = 1 = \frac{1}{2}(y-x^2) \qquad y = x^2 + 2$$

$$f = -1 = \frac{1}{2}(y-x^2) \qquad y = x^2 - 2$$

and so on as shown in Figure 11.13. These show the level curves of $f(x,y)$. Also the gradient vector ∇f is always perpendicular to the level curves pointing to the greatest increase.

$$\nabla f = \langle f_x, f_y \rangle = \langle -x, \frac{1}{2} \rangle$$

is shown with arrow heads.

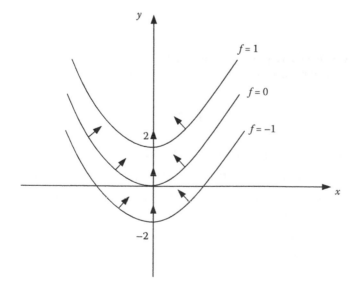

Figure 11.13 The level curves for the function.

11.4.1   Conservative Vector Fields

Why are gradient fields of importance? If $\bar{F}(x,y) = \nabla f$ for some vector field f, then $\bar{F}(x,y)$ is a *conservative vector field*.

In Figure 11.14 there is a path C within a vector field $\bar{F}(x,y)$. If $\bar{F}(x,y) = \nabla f$, then the *fundamental theorem of line integral* can be expressed by

$$\int_C \bar{F}.d\bar{r} = f(\text{end}) - f(\text{start}) \tag{11.10}$$

This is analogous to the *fundamental theorem of calculus.*

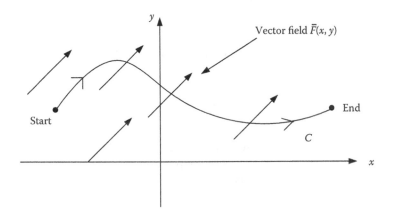

Figure 11.14 A path given within a vector field.

Example 11.12

Consider the vector field $\bar{F}(x,y) = \langle 2x + 3y, 3x \rangle$. Compute

$$\int_C \bar{F}(x,y).d\bar{r}$$

where C is the curve given in Figure 11.15.
    This could be tackled in the usual way. Work out

$$\int_C \bar{F}.d\bar{r}$$

using parameterization of the curve as before.

But since it is known that $\bar{F}(x,y) = \nabla f$, for $f(x, y) = x^2 + 3xy$, this implies that $\bar{F}$ is a conservative vector field. It is easier to use the fundamental theorem of line integrals, which gives

$$\int_C \bar{F}.d\bar{r} = f(\text{end}) - f(\text{start}) = f(3,0) - f(0,9)$$

$$= (3^2 + 0) - (0^2 + 0) = 9$$

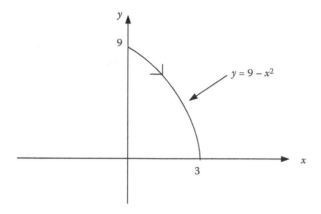

Figure 11.15 The path is the curve $y = 9 - x^2$.

Following are some important properties of the fundamental theorem of line integrals.

Property 1: If $\bar{F}$ is conservative, and C and $\tilde{C}$ have the same start and end point as shown in Figure 11.16, then it can be shown that

$$\int_C \bar{F}.d\bar{r} = \int_{\tilde{C}} \bar{F}.d\bar{r} \tag{11.11}$$

This is called *path independence* for conservative vector fields.

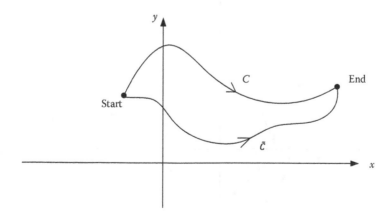

Figure 11.16 Two paths with the same start and end points.

Note: Systems that are reversible are conservative and those that are non-conservative in nature mean that energy is lost during the process as waste or to entropy.

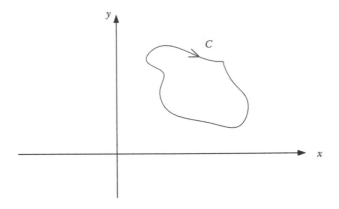

Figure 11.17 A closed curve with the same start and end points.

Property 2: If C is a closed curve (a curve that comes back on its self) as shown in Figure 11.17, then it can be shown that the integral around the *closed path* is

$$\oint_C \bar{F}.d\bar{r} = 0 \qquad \text{(conservativeness)} \qquad (11.12)$$

Can it be said that all vector fields are conservative? No, not all vector fields are conservative. Consider the following vector field (also see Figure 11.18)

$$\bar{F}(x, y) = \frac{\langle -y, x \rangle}{\sqrt{x^2 + y^2}}$$

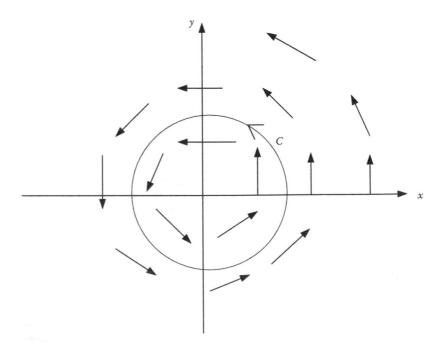

Figure 11.18 The vector field given by $\bar{F}(x, y) = \dfrac{\langle -y, x \rangle}{\sqrt{x^2 + y^2}}$.

This is not a conservative vector field since clearly

$$\oint_C \bar{F}.d\bar{r} > 0$$

11.4.2   Testing for Conservativeness

So, if

$$\bar{F}(x,y) = \langle M(x,y), N(x,y) \rangle = \nabla f$$

then we need to find some function f such that we have the following as shown in Figure 11.19.

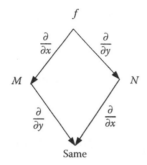

Figure 11.19 Showing that the order of differentiating does not matter.

Example 11.13

Let $\bar{F}(x,y) = \langle M(x,y), N(x,y) \rangle = \langle x^2 + y, y \rangle$ be the vector field. So,

$$\frac{\partial N}{\partial x} = N_x = 0$$

and

$$\frac{\partial M}{\partial y} = M_y = 1$$

These are not equal so the vector field $\bar{F}(x,y)$ is not conservative.

Example 11.14

Is $\bar{F}(x,y) = \langle 3x^2 + y, x \rangle$ a conservative vector field?

Check the first partial derivative tests: $N_x = 1$ and $M_y = 1$. These are necessary but not sufficient conditions. This is a good sign that maybe

$$\bar{F}(x,y) = \nabla f$$

To see if this is the case integrating the functions as follows: $f_x = 3x^2 + y$ with respect to x gives, $f = x^3 + xy + g(y)$ and $f_y = x$ with respect to y gives $f = xy + h(x)$. Now, the function f has to fit these two equations simultaneously. Therefore, it is clear that if $f(x, y) = x^3 + xy$, then this satisfies the necessary conditions and so $\overline{F}(x, y)$ is a conservative vector field.

Example 11.15

Compute the following for line integral

$$\oint_C \overline{F}.d\overline{r}$$

for $\overline{F}(x, y) = \langle 3x^2 + y, x \rangle$ along the curve C, where C is given in Figure 11.20.

Maybe $\overline{F}$ is a conservative vector field. Yes it is, as was shown in Example 11.14.

$\overline{F}(x, y) = \nabla f$, where $f(x, y) = x^3 + xy$, so the working out is easy using the fundamental theorem of line integrals:

$$\oint_C \overline{F}.d\overline{r} = f(\text{end}) - f(\text{start})$$

$$= f(4, 0) - f(0, 2) = 4^3 - 0 = 64$$

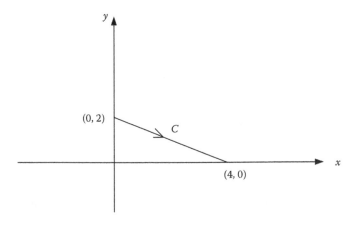

Figure 11.20 The straight-line curve C.

Notation: Given curves C_1, C_2, C_3, and C_4 as shown in Figure 11.21.
The whole curve is partitioned as $C = C_1 + C_2 + C_3 + C_4$.
Also, shown in Figure 11.22 is the negative of a curve.
The same curve as C but in the opposite direction is called $-C$.

Property 3: If $\overline{F}(x, y) = \langle M(x, y), N(x, y) \rangle$ is conservative, then $M_y = N_x$.

Does $M_y = N_x$ imply that $\bar{F}$ is conservative, that is, $\bar{F}(x, y) = \nabla f$? No, this is not always the case.

A region or domain (D) is "simply connected" if any closed curve C in D can be contracted to a point in D.

Note: Connected means that you can get from a point to another without leaving the region.

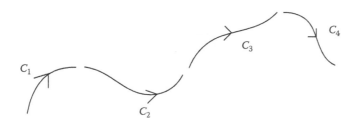

Figure 11.21 Combining different curves.

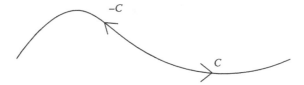

Figure 11.22 A curve and is negative.

Example 11.16

Some examples of regions D that are connected and simply connected are shown in Figure 11.23.

Theorem: If $\bar{F}(x, y) = \langle M(x, y), N(x, y) \rangle$ is defined on a *simply connected domain* and $M_y = N_x$, then it can be said that $\bar{F}$ is conservative.

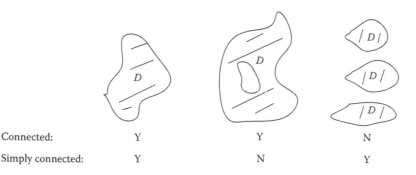

Connected:	Y	Y	N
Simply connected:	Y	N	Y

Figure 11.23 Different regions in space.

Example 11.17

An example of a famous conservative vector field is the gravitational vector force field as shown in Figure 11.24. Here the vector field is given by

$$\bar{F}(x,y) = \frac{-k\langle x,y,z\rangle}{(x^2 + y^2 + z^2)^{\frac{3}{2}}}$$

This is the gradient of f, where

$$f(x,y,z) = \frac{k}{(x^2 + y^2 + z^2)^{\frac{1}{2}}}$$

and so $\bar{F}(x,y) = \nabla f$.

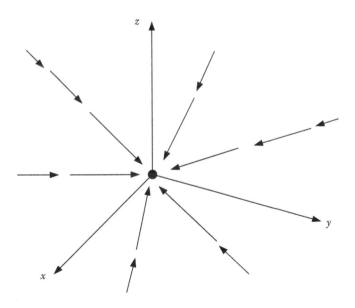

Figure 11.24 The vector field for gravity.

11.5   Green's Theorem

Having calculated line integrals for different paths as shown in Figure 11.25, using the formula

$$\int_C M(x,y)\,dx + N(x,y)\,dy$$

and having also done double integrals as follows

$$\iint_D f(x,y)\,dA$$

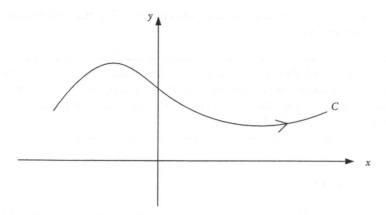

Figure 11.25 A path showing the direction of the line integral.

then Green's theorem relates these two concepts.

If the curve C is a simple, closed and positively oriented curve that surrounds a two-dimensional region D.

The following definitions are used about the curve.

Closed – The end point is the same as the start point.

Simple – Has no complexity such as no intersection with itself.

Positively oriented – As you travel along the curve C the region D is toward the left.

Pictorially, this is shown in Figure 11.26.
Then *Green's theorem* states

$$\oint_C M(x,y)\,dx + N(x,y)\,dy = \iint_D (N_x - M_y)\,dA \qquad (11.13)$$

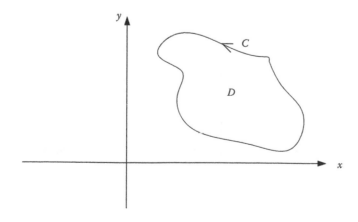

Figure 11.26 A region bounded by a closed curve.

This states that the line integral along the curve C is identical to a double integral over the region D.

Note: Just as the fundamental theorems of calculus and line integrals turn line integrals into a calculation of the function at end points, Green's theorem turns an area integral into an integral around a boundary line of the area. It is a dimension reducing method.

Example 11.18

Given $\bar{F}(x,y) = \langle M(x,y), N(x,y) \rangle = \langle y^2, xy \rangle$ is a vector field, suppose the curve C is the curve shown in Figure 11.27. Find the work done moving along C, that is, calculate the following:

$$\oint_C \bar{F}.d\bar{r}$$

Now, since $\bar{F}$ is not a conservative vector field, then it is the case that

$$\oint_C \bar{F}.d\bar{r} \neq 0$$

So, this problem can be done either directly by parameterizing the three curves and solving the $\bar{F}.d\bar{r}$ integral along each one.

Alternatively, you can use Green's theorem and change this problem into a double integral using Equation 11.13:

$$\oint_C \bar{F}.d\bar{r} = \iint_D (N_x - M_y)\,dA$$

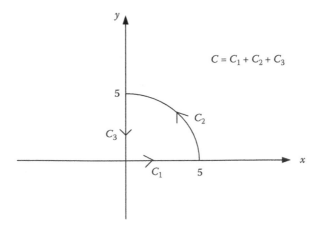

Figure 11.27 The closed curve is a combination of curves.

Here, $M(x,y) = y^2$ implies that $M_y = 2y$.
$N(x, y) = xy$ implies that $N_x = y$.

Therefore, the double integral now becomes,

$$\iint_D (N_x - M_y)\,dA = \iint_D y\,dA$$

Since the region is a quarter circle, changing to polar coordinates is better and gives

$$\iint_D (N_x - M_y)\,dA = \iint_D y\,dA = \int_0^{\frac{\pi}{2}} \int_0^5 (r\sin\theta) r\,dr\,d\theta$$

$$= \int_0^{\frac{\pi}{2}} \frac{1}{3}[r^3 \sin\theta]_0^5\,d\theta = \int_0^{\frac{\pi}{2}} \frac{125}{3}\cos\theta\,d\theta = \frac{125}{3}$$

With a closed curve, Green's theorem can be used to save some effort in the calculations.

11.5.1   Properties of Green's Theorem

Property 1: If $\bar{F}(x, y) = \langle M(x, y), N(x, y)\rangle$ is a conservative vector field, then $M_y = N_x$. But Green's theorem states

$$\oint_C \bar{F}.d\bar{r} = \iint_D (N_x - M_y)\,dA = 0$$

This ties into the previous result for conservative closed vector fields (Equation 11.12).
Property 2: The above only relies on $M_y = N_x$.

If $\bar{F}(x, y) = \dfrac{\langle -y, x\rangle}{x^2 + y^2}$ as shown in Figure 11.28, then it can be shown that $M_y = N_x$, but the vector field is not conservative.
So,

$$\int_{C_1} \bar{F}.d\bar{r} = \iint_{D_1} (N_x - M_y)\,dA = 0$$

Also, what about along C_2, since N_x and M_y are not defined at $(0, 0)$.

$$\int_{C_2} \bar{F}.d\bar{r} = \iint_{D_2} (N_x - M_y)\,dA \neq 0$$

Property 3: Where $M_y = N_x$, there is a restricted form of path independence, paths can be perturbed so long as to stay in the domain as shown in Figure 11.29.

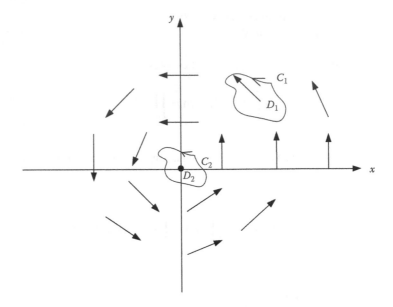

Figure 11.28 Vector field not defined at the origin.

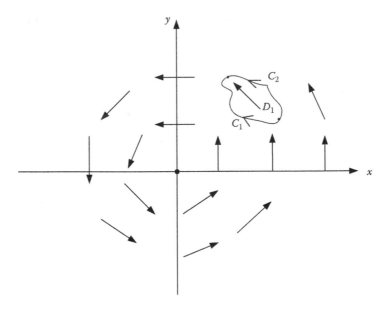

Figure 11.29 Perturbed paths in the domain.

$$\bar{F}(x,y) = \frac{\langle -y, x \rangle}{x^2 + y^2} \quad \text{with} \quad M_y = N_x$$

So,

$$\int_{C_2} \bar{F}.d\bar{r} - \int_{C_1} \bar{F}.d\bar{r} = \oint_{C_2 - C_1} \bar{F}.d\bar{r} = \iint_{D_2} (N_x - M_y)\, dA = 0$$

This implies that

$$\int_{C_1} \overline{F}.d\overline{r} = \int_{C_2} \overline{F}.d\overline{r}$$

Property 4: Green's theorem relates a 2-D region D to its 1-D boundary C, as shown in Figure 11.30.

A similar concept is the fundamental theorem of line integrals, which relates a 1-D curve to its 0-D boundary points, as shown in Figure 11.31.

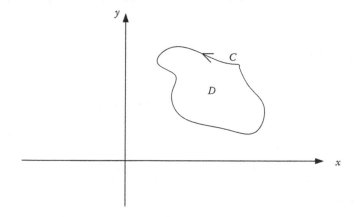

Figure 11.30 A region with a closed boundary.

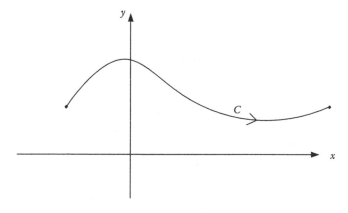

Figure 11.31 1-D curve with 0-D boundary points.

11.6 Divergence and Curl of Vector Fields

11.6.1 2-Dimensional Definitions

In 2-D space if $\overline{F}(x, y) = \langle M(x, y), N(x, y) \rangle$ is a vector field, then the following definitions are used.

$$\text{div } \overline{F} = M_x + N_y \quad \text{is called the divergence of } \overline{F}. \quad (11.14)$$

$$\text{curl } \bar{F} = N_x - M_y \quad \text{is called the curl of } \bar{F}. \tag{11.15}$$

How do we interpret these quantities? It first helps to think of $\bar{F}(x,y)$ as a velocity field of a flow of water. Then,

Divergence – Measures the net flow out a point.

Curl – Measures the counterclockwise (ccw) rotation at a point. This implies that curl is positive if there is a ccw rotation."

Now let's look at some examples that illustrate these concepts with vector fields.

Example 11.19

A vector field is given by $\bar{F}(x,y) = \langle x, y \rangle$ and shown in Figure 11.32. Consider the divergence at any point. Place a tiny box at a point and asking if there is more flow into the box or more flow coming out of the box. Here it is the case that there should be more flow out of a box at a point then flows in, that is, div $\bar{F} > 0$.

For the curl of this field, consider placing a paddle wheel into the field and see if it rotates. The paddle wheel will flow outward but does not rotate, that is, curl $\bar{F} = 0$.

So, computing the divergence and curl of $\bar{F}$ using the formulae given by Equations 11.14 and 11.15 gives

$$\text{div } \bar{F} = M_x + N_y = 1 + 1 = 2 > 0$$

$$\text{curl } \bar{F} = N_x - M_y = 0 + 0 = 0$$

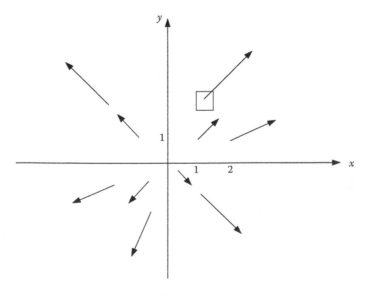

Figure 11.32 Vector field given by $\bar{F}(x,y) = \langle x, y \rangle$.

Example 11.20

Given the vector field, $\bar{F}(x,y) = \langle y, 0 \rangle$, as shown in Figure 11.33, calculating the divergence and curl gives

$$\text{div } \bar{F} = M_x + N_y = 0 + 0 = 0$$

$$\text{curl } \bar{F} = N_x - M_y = 0 - 1 = -1 \quad \text{(i.e., an anticlockwise rotation)}$$

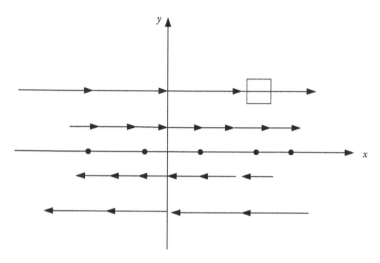

Figure 11.33 Vector field given by $\bar{F}(x,y) = \langle y, 0 \rangle$.

Example 11.21

In a vector field, $\bar{F}(x,y) = \langle y, xy \rangle$. This is a complicated vector field to see easily, but it is still possible to compute the divergence and curl of $\bar{F}$ using the definitions.

$$\text{div } \bar{F} = M_x + N_y = 0 + x = x$$

$$\text{curl } \bar{F} = N_x - M_y = y - 1 = y - 1$$

These depend on the location in the x-y plane.

11.6.2   Alternative Forms of Green's Theorem

Now having the definitions of the divergence and curl of a vector field, it is possible to write Green's theorem using these forms. The region is D and is bounded by a curve C, as shown in Figure 11.34.

11.6.2.1   Curl Form

Using the definition of Green's theorem given by Equation 11.13,

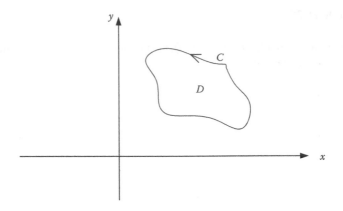

Figure 11.34 Region bounded by a curve.

$$\oint_C M(x,y)\,dx + N(x,y)\,dy = \iint_D (N_x - M_y)\,dA$$

This can now also be written as using the definition of the curl:

$$\int_C \bar{F}.\,d\bar{r} = \iint_D \operatorname{curl}\bar{F}\,dA \qquad (11.16)$$

So, the curl $\bar{F}$ represents how much counterclockwise rotation there is at each point and if all of these are added up, this then gives the total rotation around the boundary.

Also note that now if $\bar{F}$ is a conservative ($M_y = N_x$), then this implies that curl $\bar{F} = 0$.

11.6.2.2   Divergence Form

Figure 11.35 shows a bounded region within a vector field.

$$\oint_C (\bar{F}.\hat{n})\,ds = \oint_C M\,dy - N\,dx$$

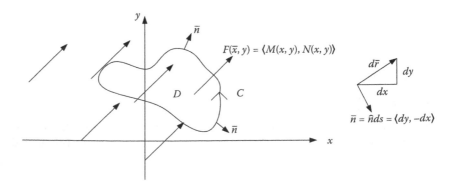

Figure 11.35 A bounded region within a vector field.

But using Green's theorem gives

$$\oint_C M\,dy - N\,dx = \iint_D (M_x - -N_y)\,dA = \iint_D (M_x + N_y)\,dA = \iint_D \text{div}\,\bar{F}\,dA$$

So, this gives the following result:

$$\oint_C (\bar{F}.\hat{n})\,ds = \iint_D \text{div}\,\bar{F}\,dA \qquad (11.17)$$

This shows that the net flow out at each point all added up together gives the total flux out of the curve shown in Figure 11.35.

11.6.3   3-Dimensional Definitions

Given a vector field in 3-D space, $\bar{F}(x,y,z) = \langle M(x,y,z), N(x,y,z), P(x,y,z) \rangle$, then the following are the definitions for the divergence and curl of the vector field:

$$\text{div}\,\bar{F} = M_x + N_y + P_z \qquad \text{is called the divergence of } \bar{F} \text{ in 3-D.} \qquad (11.18)$$

$$\text{curl}\,\bar{F} = \langle P_y - N_z, M_z - P_x, N_x - M_y \rangle \text{ is called the curl of } \bar{F} \text{ in 3-D and is a vector.}$$

$$(11.19)$$

It is easier to remember the definitions for grad $\bar{F}$, div $\bar{F}$, and the curl $\bar{F}$ in terms of the "del or nabla notation."
Let

$$\nabla \triangleq \langle \frac{\partial}{\partial x}, \frac{\partial}{\partial y}, \frac{\partial}{\partial z} \rangle$$

then the following can be defined:

$$\text{grad}\,\bar{F} = \nabla \bar{F} \qquad (11.20)$$

$$\text{div}\,\bar{F} = \nabla.\bar{F} \qquad (11.21)$$

$$\text{curl}\,\bar{F} = \nabla \times \bar{F} \qquad (11.22)$$

Note: The div $\bar{F}$ still measures the net flow out at a point. However, the curl $\bar{F}$ is more difficult to see. Generally, the curl $\bar{F}$ points in the direction of the "broom handle" with the largest counterclockwise rotation of the paddle wheel. The magnitude of curl $\bar{F}$ is the amount of counterclockwise rotation.

Example 11.22

Given the vector field,

$$\bar{F}(x,y,z) = \langle M(x,y,z), N(x,y,z), P(x,y,z) \rangle = \langle x^2, xy, z \rangle,$$

find the divergence and curl of the vector field. Using the definitions gives

$$\text{div } \bar{F} = M_x + N_y + P_z = 2x + x + 1 = 3x + 1$$

$$\text{curl } \bar{F} = \nabla \times \bar{F} = \begin{vmatrix} i & j & k \\ \dfrac{\partial}{\partial x} & \dfrac{\partial}{\partial y} & \dfrac{\partial}{\partial z} \\ x^2 & xy & z \end{vmatrix}$$

$$= \langle 0 - 0, 0 - 0, y - 0 \rangle$$

$$= \langle 0, 0, y \rangle$$

Some facts can now be stated as follows:

1. (2-D) curl $(\nabla f) = 0$ alternative notation gives $\nabla \times \nabla f = 0$.
2. (3-D) curl $(\nabla f) = \bar{0}$ alternative notation gives $\nabla \times \nabla f = \bar{0}$.
3.   div(curl $\bar{F}$) $= 0$ alternative notation gives $\nabla . \nabla \times \bar{F} = 0$.

Proofs:

1. $\nabla f = \langle f_x, f_y \rangle$,   then   curl $(\nabla f) = f_{yx} - f_{xy} = 0$.
2. $\nabla f = \langle f_x, f_y, f_z \rangle$, then

$$\text{curl } (\nabla f) = \begin{vmatrix} i & j & k \\ \dfrac{\partial}{\partial x} & \dfrac{\partial}{\partial y} & \dfrac{\partial}{\partial z} \\ f_x & f_y & f_z \end{vmatrix}$$

$$= \langle f_{zy} - f_{yz}, f_{xz} - f_{zx}, f_{yx} - f_{xy} \rangle = \langle 0, 0, 0 \rangle = \bar{0}$$

3. $\operatorname{curl} \bar{F} = \nabla \times \bar{F} = \begin{vmatrix} \underline{i} & \underline{j} & \underline{k} \\ \dfrac{\partial}{\partial x} & \dfrac{\partial}{\partial y} & \dfrac{\partial}{\partial z} \\ M & N & P \end{vmatrix} = \langle P_y - N_z, M_z - P_x, N_x - M_y \rangle$

Therefore, $\operatorname{div}(\operatorname{curl} \bar{F}) = P_{yx} - N_{zx} + M_{zy} - P_{xy} + N_{xz} - M_{yz} = 0$.

Note: $\nabla . \nabla \times \bar{F} = 0$. *The* $\nabla \times \bar{F}$ *is a vector perpendicular to both* ∇ *and* $\bar{F}$, *so its dot product with* ∇ *should be zero.*

11.7 Surface Integration

11.7.1 Parametric Surfaces

For a line there was one parameter. Generally for surfaces there are more parameters starting with two, say, u and v.

$$x = x(u, v), \quad y = y(u, v), \quad z = z(u, v), \quad \text{where } (u, v) \in D$$

Or these can be written as

$$\bar{r}(u, v) = \langle x(u, v), y(u, v), z(u, v) \rangle, \quad \text{where } (u, v) \in D$$

Pictorially, this can be shown as in Figure 11.36.

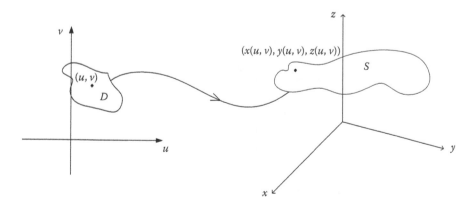

Figure 11.36 Region maps unto a surface.

Example 11.23

A graph of a function in terms of an equation given the surface $z = 4 - x^2 - y^2$ and $z \geq 0$ is shown in Figure 11.37.

To parameterize the surface, let $x = u$, $y = v$, then $z = 4 - u^2 - v^2$, where $(u, v) \in D$ (Figure 11.38).

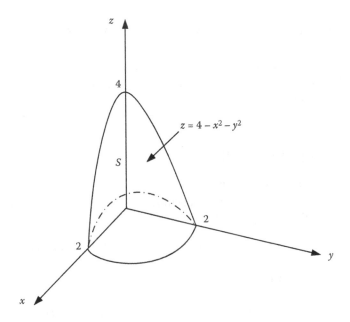

Figure 11.37 Region bounded by a surface.

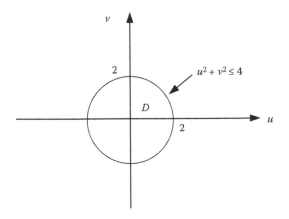

Figure 11.38 The region on the x-y plane.

Example 11.24

A spherical shell of radius 5 is shown in Figure 11.39. In spherical coordinates, this a sphere with radius $\rho = 5$.

So letting $u = \theta$ and $v = \varphi$, then a logical parameterization is

$$x = 5\sin v \cos u$$

$$y = 5\sin v \sin u$$

$$z = 5\cos v$$

$(u, v) \in D$ and the limits on u and v are shown in Figure 11.40.

Note: The sphere in 3D (x, y, z) is mapped to a rectangle in 2D (u, v).

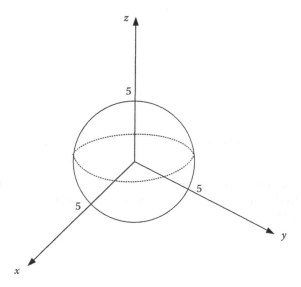

Figure 11.39 A spherical shell of radius 5.

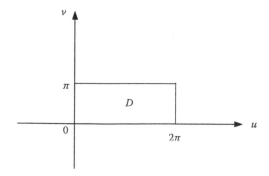

Figure 11.40 The limits of the region in the *u-v* plane.

11.7.1.1   *Summary of the Main Types of Surfaces*

1. The graph of the function $z = f(x, y)$ is shown in Figure 11.41.

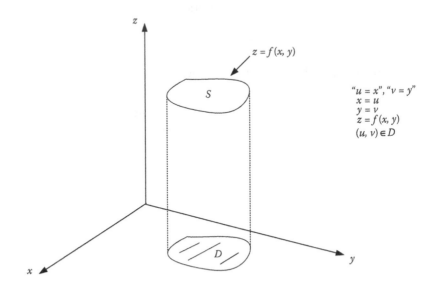

Figure 11.41 Graph of a function.

2. In some coordinate systems, one of the coordinates is constant (e.g., in cylindrical coordinates), the radius r = constant = R as shown in Figure 11.42.

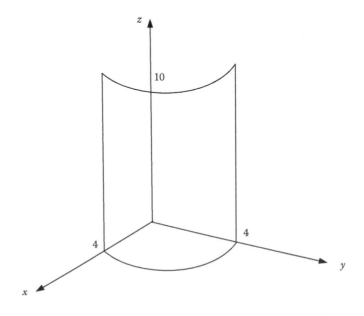

Figure 11.42 Fixing a coordinate in a system.

To parameterize this surface S, in this case R is fixed at 4. Then using cylindrical coordinates gives $x = 4 \cos u$, $y = 4\sin u$, and $z = v$. Here $u = \theta$ and $v = z$, and

$$0 \le u \le \frac{\pi}{2} \quad \text{and} \quad 0 \le v \le 10$$

The need to parameterize surfaces leads to surface integrals.

11.7.2   Surface Integrals

The idea of a surface is similar to a region as shown in Figure 11.43.

Splitting the surface into small pieces and summing gives the definition of a surface integral as

$$\iint_S f(x, y, z)\, dS \triangleq \lim_{\Delta S \to 0} \sum_{all\ pieces} f(x^*, y^*, z^*)\Delta S \qquad (11.23)$$

Usual applications of surface integrals are

$$\text{Area}\,(S) = \iint_S 1\, dS \qquad (11.24)$$

$$\text{Mass}\,(S) = \iint_S d(x, y, z)\, dS; \qquad (11.25)$$

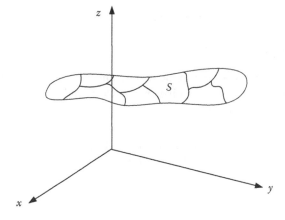

Figure 11.43 A general surface in 3-D space.

where d is the density of a piece.

There is more work to do here when computing surface integrals. This can be thought of as a "change of variables," as shown in Figure 11.44.

Using the definition of transforming variables gives

$$\iint_S g(x, y, z)\, dS = \iint_D g[x(u, v), y(u, v), z(u, v)] \left| \frac{\partial(x, y, z)}{\partial(u, v)} \right| dv\, du \qquad (11.26)$$

Now, the Jacobian is $\dfrac{\partial(x,y,z)}{\partial(u,v)} = \det\begin{pmatrix} x_u & x_v \\ y_u & y_v \\ z_u & z_v \end{pmatrix}$

The generalized determinant of a rectangular matrix is given by

$$\det\begin{pmatrix} x_u & x_v \\ y_u & y_v \\ z_u & z_v \end{pmatrix} \triangleq \sqrt{\det\begin{pmatrix} x_u & x_v \\ y_u & y_v \end{pmatrix}^2 + \det\begin{pmatrix} x_u & x_v \\ z_u & z_v \end{pmatrix}^2 + \det\begin{pmatrix} y_u & y_v \\ z_u & z_v \end{pmatrix}^2} \quad (11.27)$$

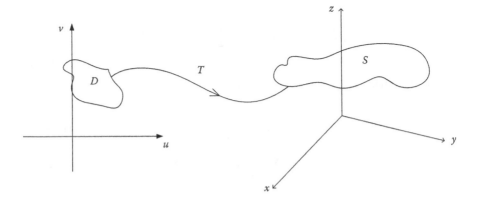

Figure 11.44 A transformation of variables.

Example 11.25

Find the area of $x + y + z = 1$ in the first octant as shown in Figure 11.45.

Parameterizing the surface gives $x = u$, $y = v$, and $z = 1 - u - v$. So, the area is given by the formula

$$\text{Area}(S) = \iint_S 1\, dS = \iint_D 1 \left| \frac{\partial(x,y,z)}{\partial(u,v)} \right| dv\, du$$

$$\frac{\partial(x,y,z)}{\partial(u,v)} = \det\begin{pmatrix} x_u & x_v \\ y_u & y_v \\ z_u & z_v \end{pmatrix} = \sqrt{\det\begin{pmatrix} 1 & 0 \\ 0 & 1 \end{pmatrix}^2 + \det\begin{pmatrix} 1 & 0 \\ -1 & -1 \end{pmatrix}^2 + \det\begin{pmatrix} 0 & 1 \\ -1 & -1 \end{pmatrix}^2}$$

$$= \sqrt{(1)^2 + (-1)^2 + (1)^2} = \sqrt{3}$$

$$\text{Area}(S) = \sqrt{3} \iint_D dv\, du$$

$$= \sqrt{3}\, \text{Area}(D) = \frac{\sqrt{3}}{2}$$

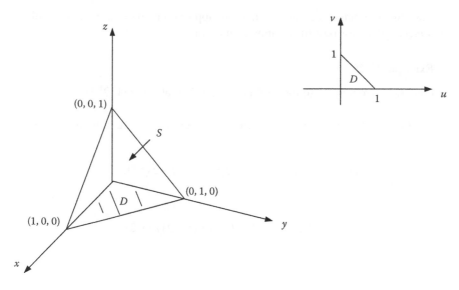

Figure 11.45 Region in the first quadrant.

11.7.3   Tangent Planes and Normal Vectors

Graphically, Figure 11.46 shows a surface and a tangent plane to it. (x, y) is a point near (a, b) but using the Taylor series expansion gives

$$f(x,y) \approx f(a,b) + (x-a)f_x(a,b) + (y-b)f_y(a,b)$$

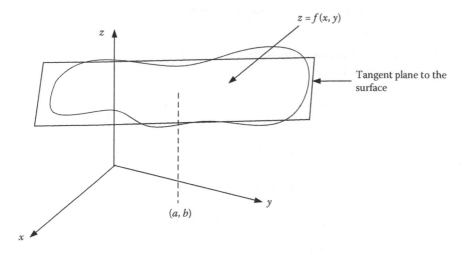

Figure 11.46 A surface given by $z = f(x, y)$ and its tangent plane.

Call the right-hand side equal to z, then this can be written as

$$z = f(a,b) + (x-a)f_x(a,b) + (y-b)f_y(a,b) \qquad (11.28)$$

This is an approximation to $f(x, y)$.

Now, this is a plane accurate at (a, b) but approximate near (a, b) and is called the *tangent plane* (or best linear approximation).

Example 11.26

Find the tangent plane to the surface $z = x + 5xy^2$ at point (1, 2, 3).

Solution: Let the function $f(x, y) = x + 5xy^2$. Therefore, using Equation 11.28 gives

$$f(x, y) \approx f(1, 2) + (x - 1)f_x(1, 2) + (y - 2)f_y(1, 2)$$

Let

$$z = f(1, 2) + (x - 1)f_x(1, 2) + (y - 2)f_y(1, 2)$$

$$f(1, 2) = 21$$

$$f_x = 1 + 5y^2 \quad \rightarrow \quad f_x(1, 2) = 21$$

$$f_y = 10xy \quad \rightarrow \quad f_y(1, 2) = 20$$

$$z = 21 + 21(x - 1) + 20(y - 2)$$

$21x + 20y - z = 40$ is the equation of the required tangent plane.

11.7.3.1   Normal Vector to the Tangent Plane
Example 11.27

Find the normal vector to the surface $z = x^2 + y^2$ at the point (−1, 2, 2), as shown in Figure 11.47.

Solution: $\underline{n}$ is a normal vector perpendicular to the surface at (−1, 2, 2). First, finding the tangent plane using $f(x, y) = x^2 + y^2$ with Equation 11.28 gives

$$z = f(-1, 2) + (x + 1)f_x(-1, 2) + (y - 2)f_y(-1, 2)$$

$$f(x, y) = x^2 + y^2 \quad f(-1, 2) = 5$$

$$f_x = 2x \quad \rightarrow \quad f_x(-1, 2) = -2$$

$$f_y = 2y \quad \rightarrow \quad f_y(-1, 2) = 4$$

$$z = 5 - 2(x + 1) + 4(y - 2)$$

$$2x - 4y + z = -5$$

This is the tangent plane at (−1, 2, 2).

But the normal to the surface is the normal to the tangent plane. So, the normal vector to the tangent plane $2x - 4y + z = -5$ is just $\underline{n} = \langle 2, -4, 1 \rangle$

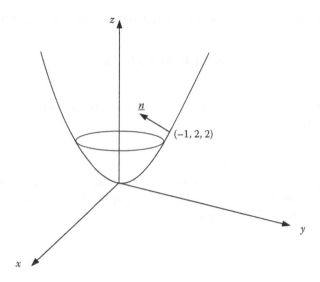

Figure 11.47 Surface and its normal vector at the given point.

11.7.3.2   General Formula for the Normal Vector to a Surface

In Figure 11.48 is a general surface $z = f(x, y)$ and its tangent plane. The tangent plane is given by Equation 11.28 as

$$z = f(a,b) + (x - a)f_x(a,b) + (y - b)f_y(a,b)$$

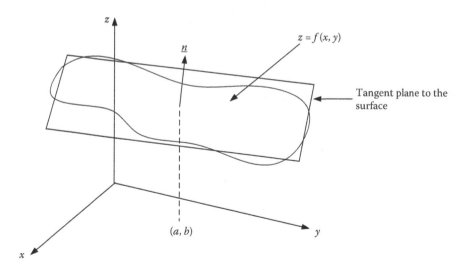

Figure 11.48 General surface and its normal vector.

This can be rewritten as x, y, and z terms on one side with all the constants on the other side as

$$-f_x(a,b)x - f_y(a,b)y + z = \text{some constant}$$

Therefore, the normal vector also normal to the surface at the given point is

$$\underline{n} = \langle -f_x(a,b). - f_y(a,b), 1 \rangle \qquad (11.29)$$

This is pointing up as $z = 1$ and this formula will be useful in later sections on surface integrals.

11.7.4   Normal Vectors to Surfaces

Normal vectors to surfaces can be pointing in different directions as shown in Figure 11.49. Normal vectors can be pointing upward or outward as shown in Figures 11.50 and 11.51, respectively.

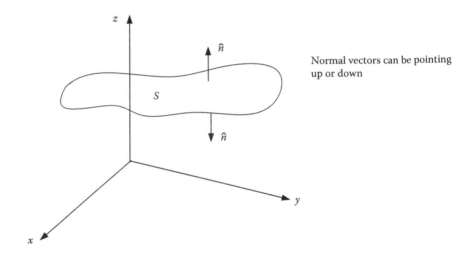

Normal vectors can be pointing up or down

Figure 11.49 Normal vectors to a surface.

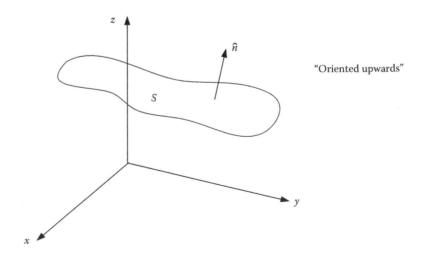

"Oriented upwards"

Figure 11.50 Normal vector pointing upward.

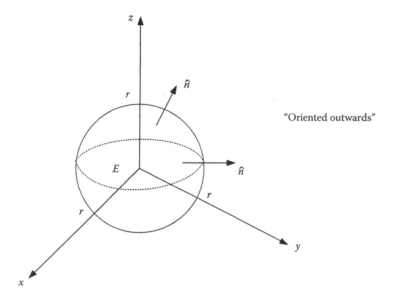

Figure 11.51 Normal vector pointing outward.

There are different ways to find the normal vectors to different surfaces. For a general graph as shown in Figure 11.52 the normal vector is simply given by using the formula for $z = f(x,y)$ as

$$\hat{n} = \frac{\langle -f_x, -f_y, 1 \rangle}{\sqrt{f_x^2 + f_y^2 + f_z^2}} \qquad (11.30)$$

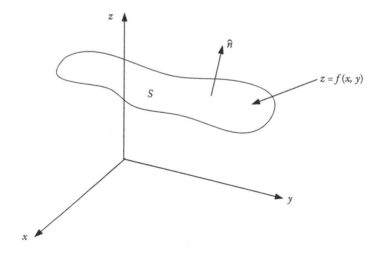

Figure 11.52 Normal vector for a surface given by a graph.

The case of a sphere is shown in Figure 11.53. The normal vector is given by using the formula given by Equation 11.29 as

$$\hat{n} = \frac{\langle x, y, z \rangle}{\sqrt{x^2 + y^2 + z^2}} \qquad (11.31)$$

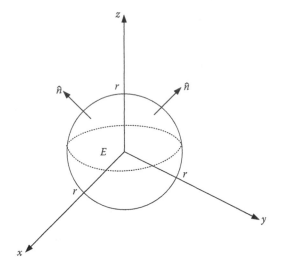

Figure 11.53 Normal vectors to a sphere.

In the case of a cylindrical surface is shown in Figure 11.54. The normal vector is given by using the formula given in Equation 11.31 as

$$\hat{n} = \frac{\langle x, y, 0 \rangle}{\sqrt{x^2 + y^2}} \tag{11.32}$$

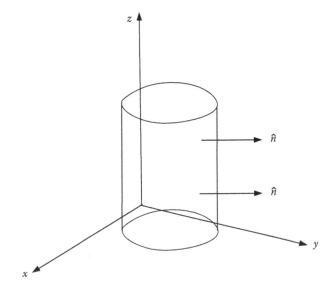

Figure 11.54 Normal vectors to a cylinder.

Generally, for a surface of the form $\bar{r} = (u, v)$, the normal vector is

$$\hat{n} = \frac{\bar{r}_u \times \bar{r}_v}{|\bar{r}_u \times \bar{r}_v|} \tag{11.33}$$

Note: This formula can always be used if required.

11.7.5   Applications of Surface Integrals

The flux of a vector $\bar{F}$ across a surface S is shown in Figure 11.55. The flux of a vector $\bar{F}$ across the surface element ΔS is calculated by considering the component of $\bar{F}$ perpendicular to the surface, that is, in the direction of $\hat{n}$ as $(\bar{F}.\hat{n})$. The flux coming out of the surface element ΔS is then given by $(\bar{F}.\hat{n})\Delta S$.

The total flux coming out of the surface S can be found by a summing of $(\bar{F}.\hat{n})\Delta S$ as $\Delta S{\rightarrow}0$ which then produces a double integral:

$$\text{Flux across the surface} = \iint_S (\bar{F}.\hat{n})\,dS \qquad (11.34)$$

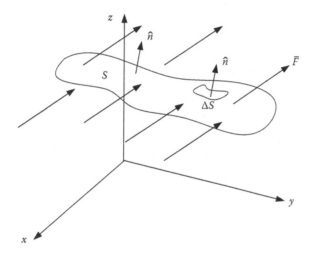

Figure 11.55 Flux across a surface.

Example 11.28

Compute the flux of the vector field $\bar{F} = \langle x, y, z \rangle$ across the surface S, where S is given by $z = 4 - x^2 - y^2$ and $z \geq 0$ and oriented upward as shown in Figure 11.56.

Solution: The normal vector $\hat{n}$ is given by the formula $\hat{n} = \dfrac{\langle -f_x, -f_y, 1 \rangle}{\sqrt{f_x^2 + f_y^2 + f_z^2}}$, which gives

$$\hat{n} = \frac{\langle 2x, 2y, 1 \rangle}{\sqrt{4x^2 + 4y^2 + 1}}$$

$$\text{Flux} = \iint_S (\bar{F}.\hat{n})\,dS = \iint_S \langle x, y, z \rangle . \frac{\langle 2x, 2y, 1 \rangle}{\sqrt{4x^2 + 4y^2 + 1}}\,dS$$

$$= \iint_S \frac{2x^2 + 2y^2 + z}{\sqrt{4x^2 + 4y^2 + 1}}\,dS$$

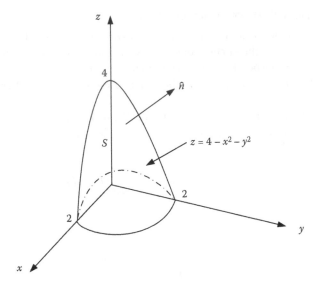

Figure 11.56 Flux out of a surface.

Now to parameterize the surface S, let $x = u$, $y = v$, and $z = 4 - u^2 - v^2$, where $(u, v) \in D$ (see Figure 11.57):

$$= \iint_S \frac{2x^2 + 2y^2 + z}{\sqrt{4x^2 + 4y^2 + 1}} dS = \iint_D \frac{(2u^2 + 2v^2) + (4 - u^2 - v^2)}{\sqrt{4u^2 + 4v^2 + 1}} \left| \frac{\partial(x, y, z)}{\partial(u, v)} \right| dv\, du$$

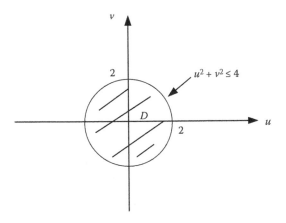

Figure 11.57 Region in the u-v plane.

$$\frac{\partial(x, y, z)}{\partial(u, v)} = \det \begin{pmatrix} x_u & x_v \\ y_u & y_v \\ z_u & z_v \end{pmatrix}$$

$$= \det \begin{pmatrix} 1 & 0 \\ 0 & 1 \\ -2u & -2v \end{pmatrix} = \sqrt{(1)^2 + (2v)^2 + (2u)^2} = \sqrt{1 + 4u^2 + 4v^2}$$

$$= \iint_S \frac{2x^2 + 2y^2 + z}{\sqrt{4x^2 + 4y^2 + 1}} \, dS = \iint_D \frac{(4 + u^2 + v^2)}{\sqrt{4u^2 + 4v^2 + 1}} \sqrt{1 + 4u^2 + 4v^2} \, dv \, du$$

$$= \iint_D (4 + u^2 + v^2) \, dv \, du$$

This double integral can be performed using polar coordinates:

$$\int_0^{2\pi} \int_0^2 (4 + r^2) r \, dr \, d\theta = 24\pi$$

Example 11.29

Find the surface area of the cylinder shown in Figure 11.58.
    Area is given by the formula

$$\text{Area}(S) = \iint_S 1 \, dS = \int_0^{2\pi} \int_0^H 1 \left| \frac{\partial(x, y, z)}{\partial(\theta, z)} \right| dz \, d\theta$$

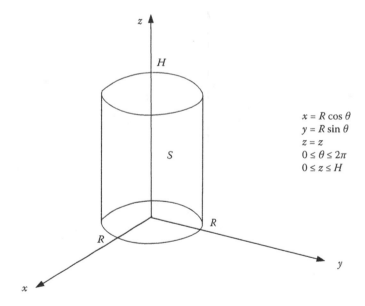

$$
\begin{aligned}
x &= R \cos\theta \\
y &= R \sin\theta \\
z &= z \\
0 &\le \theta \le 2\pi \\
0 &\le z \le H
\end{aligned}
$$

Figure 11.58 Cylinder of height H and radius R.

Now,

$$\frac{\partial(x, y, z)}{\partial(\theta, z)} = \det \begin{pmatrix} x_\theta & x_z \\ y_\theta & y_z \\ z_\theta & z_z \end{pmatrix} = \det \begin{pmatrix} -R \sin\theta & 0 \\ R \cos\theta & 0 \\ 0 & 1 \end{pmatrix}$$

$$= \sqrt{(0)^2 + (-R\sin\theta)^2 + (R\cos\theta)^2}$$

$$\frac{\partial(x,y,z)}{\partial(\theta,z)} = R$$

$$= \int\limits_{0}^{2\pi} \int\limits_{0}^{H} R\,dz\,d\theta$$

$$= \int\limits_{0}^{2\pi} RH\,d\theta = 2\pi RH$$

Example 11.30

Compute the flux out of a sphere given by $x^2 + y^2 + z^2 = 4$ for the vector field given by $\bar{F} = \langle x,y,z \rangle$ and shown in Figure 11.59.

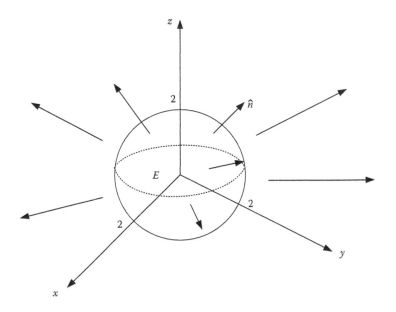

Figure 11.59 Flux out of a sphere.

$$\text{Flux} = \iint\limits_{S} (\bar{F}.\hat{n})\,dS = \iint\limits_{S} \langle x,y,z \rangle . \frac{\langle x,y,z \rangle}{\sqrt{x^2+y^2+z^2}}\,dS$$

$$= \iint\limits_{S} \frac{x^2+y^2+z^2}{\sqrt{x^2+y^2+z^2}}\,dS = \iint\limits_{S} \sqrt{x^2+y^2+z^2}\,dS$$

Parameterizing the surface with $(p = 2)$ gives

$$x = 2\sin\varphi\cos\theta$$
$$y = 2\sin\varphi\cos\theta$$
$$z = 2\cos\varphi$$

$0 \le \theta \le 2\pi$ and $0 \le \varphi \le \pi$

$$= \int_0^{2\pi}\int_0^{\pi} \rho \left|\frac{\partial(x,y,z)}{\partial(\varphi,\theta)}\right| d\varphi\, d\theta$$

$$\frac{\partial(x,y,z)}{\partial(\varphi,\theta)} = \det\begin{pmatrix} x_\varphi & x_\theta \\ y_\varphi & y_\theta \\ z_\varphi & z_\theta \end{pmatrix} = \det\begin{pmatrix} 2\cos\varphi\cos\theta & -2\sin\varphi\sin\theta \\ 2\cos\varphi\sin\theta & 2\sin\varphi\cos\theta \\ 2\sin\varphi & 0 \end{pmatrix} = 4\sin\varphi$$

$$= \int_0^{2\pi}\int_0^{\pi} 2(4\sin\varphi)\, d\varphi\, d\theta = \int_0^{2\pi} 16\, d\theta = 32\pi$$

11.8 Stokes' Theorem

There are many theorems that relate the integral over a region to the integral over the boundary of the region. Some that have already been considered are the following.

In 1-D, Figure 11.60 shows a line interval from [a, b].

Figure 11.60 Line interval from $[a,b]$.

$$\int_a^b f'(x)\, dx = f(b) - f(a)$$

This is the fundamental theorem of calculus.

In 2-D or 3-D, see Figure 11.61.

For conservative vector fields, the work done along a curve as shown in Figure 11.61 is

$$\int_C \nabla f . d\bar{r} = f(\text{end}) - f(\text{start})$$

This is the fundamental theorem of line integrals.

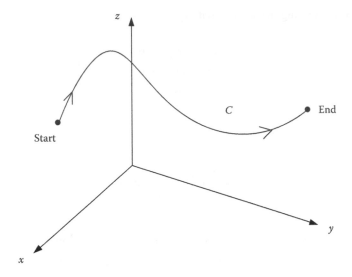

Figure 11.61 A line path in space.

In 2-D, Figure 11.62 shows a bounded region in 2-D space.

$$\iint_D \text{curl}\overline{F}\, dA = \int_C \overline{F}.d\overline{r}$$

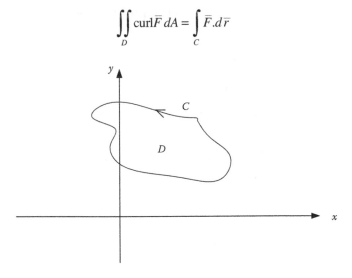

Figure 11.62 Region bounded by a curve.

This is Green's theorem. This can be extended to a 3-D surface and its boundary, as shown in Figure 11.63.

Suppose C is a simple, closed and positively oriented curve with respect to the surface S as earlier. *Stokes' theorem* states

$$\iint_S (\text{curl } \overline{F} . \, \hat{n})\, dS = \int_C \overline{F}.d\overline{r} \qquad (11.35)$$

Flux of the curl of F

An idea of what is being represented is shown in Figure 11.64.

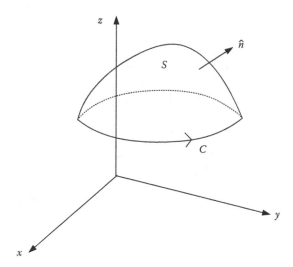

Figure 11.63 A surface in 3-D surface bounded by a curve.

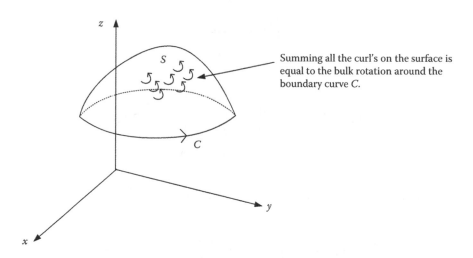

Summing all the curl's on the surface is equal to the bulk rotation around the boundary curve C.

Figure 11.64 Pictorial representation of Stokes' theorem.

Example 11.31

Verify Stokes' theorem for the vector field $\bar{F} = \langle x, z, -y \rangle$ and the surface S, where S is defined by $x + y + z = 1$ and $x^2 + y^2 \leq 1$.

The surface S is given by the plane being effectively chopped by cylinder of radius 1 as shown in Figure 11.65.

To verify Stokes' theorem the following has to be shown:

$$\iint_S (\text{curl}\bar{F} . \hat{n})\, dS = \int_C \bar{F} . d\bar{r} \qquad (11.36)$$

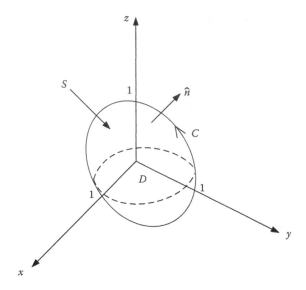

Figure 11.65 Surface in 3-D space.

Starting with the left-hand side of Equation 11.36 gives

$$\iint_S (curl\overline{F}.\hat{n})\, dS$$

$$curl\ \overline{F} = \nabla \times \overline{F} = \begin{vmatrix} \underline{i} & \underline{j} & \underline{k} \\ \dfrac{\partial}{\partial x} & \dfrac{\partial}{\partial y} & \dfrac{\partial}{\partial z} \\ x & z & -y \end{vmatrix} = \langle -2,0,0 \rangle$$

A unit normal vector to the surface is given by $\hat{n} = \dfrac{\langle 1,1,1 \rangle}{\sqrt{3}}$

$$\iint_S (curl\overline{F}.\hat{n})\, dS = \iint_S \langle -2,0,0 \rangle . \dfrac{\langle 1,1,1 \rangle}{\sqrt{3}}\, ds = \iint_S \dfrac{-2}{\sqrt{3}}\, dS$$

Parameterize the surface S,

$$x = x$$

$$y = y$$

$$z = 1 - x - y$$

where $(x, y) \in D$

$$= \iint_D \dfrac{-2}{\sqrt{3}} \left| \dfrac{\partial(x,y,z)}{\partial(x,y)} \right| dy\, dx$$

$$\frac{\partial(x,y,z)}{\partial(x,y)} = \det\begin{pmatrix} x_x & x_y \\ y_x & y_y \\ z_x & z_y \end{pmatrix} = \det\begin{pmatrix} 1 & 0 \\ 0 & 1 \\ -1 & -1 \end{pmatrix} = \sqrt{(1)^2 + (-1)^2 + (-1)^2} = \sqrt{3}$$

$$= \iint_D -2\,dy\,dx = -2\,\text{Area}(D) = -2\pi$$

The right-hand side of the Equation 11.37 gives

$$\int_C \bar{F}.d\bar{r} = \int_C x\,dx + z\,dy + (-y)\,dz = \int_C x\,dx + z\,dy - y\,dz$$

where C is the boundary curve, which is the intersection of the plane $x + y + z = 1$ and the circle $x^2 + y^2 = 1$.

$$x = \cos t$$
$$y = \sin t$$
$$z = 1 - \cos t - \sin t$$

$0 \le t \le 2\pi$

$$= \int_C x\,dx + z\,dy - y\,dz$$

$$= \int_0^{2\pi} [(\cos t)(-\sin t) + (1 - \cos t - \sin t)(\cos t) - (\sin t)(\sin t - \cos t)]\,dt$$

$$= \int_0^{2\pi} (-1 + \cos t - \cos t \sin t)\,dt$$

Using trigonometric identities and properties of the cosine and sine functions gives

$$= \int_0^{2\pi} \left(-1 + \cos t - \frac{1}{2}\sin 2t\right)dt = -2\pi$$

So the left-hand side equals the right-hand side and Stokes' theorem is verified.

11.9   Divergence Theorem

In Figure 11.66 there is a 3-D region (E) bounded by a 2-D surface-oriented outward. If all the flux at each point in the region (E) are added together, as shown in Figure 11.67, then this will give the net flux out of the surface. Mathematically, this is stated using the *divergence theorem* as

$$\iiint_E \operatorname{div} \bar{F}\, dV = \iint_S (\bar{F} \cdot \hat{n})\, dS \tag{11.37}$$

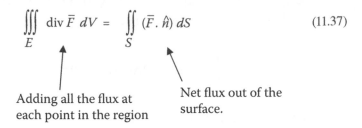

Adding all the flux at
each point in the region

Net flux out of the
surface.

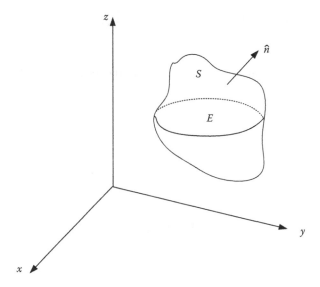

Figure 11.66 3-D region bounded by a 2-D surface.

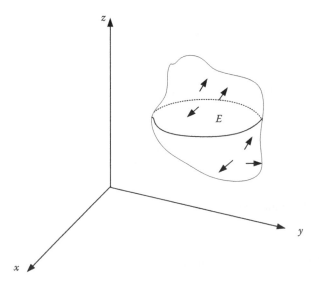

Figure 11.67 Adding all flux at each point in the region.

Example 11.32

Verify the divergence theorem for the vector field $\bar{F} = \langle x, y, z \rangle$ for the following sphere of radius 1 and with a boundary surface S as shown in Figure 11.68.

Solution: Starting with the definition, show the following:

$$\iiint_E \text{div}\bar{F}\,dV = \iint_S (\bar{F}.\hat{n})\,dS \qquad (11.38)$$

Starting with the left-hand side of Equation 11.38 gives

$$\text{div } \bar{F} = M_x + N_y + P_z = 1 + 1 + 1 = 3$$

$$\iiint_E \text{div}\bar{F}\,dv = \iiint_E 3\,dV = 3\text{Vol}(E) = 3\frac{4}{3}\pi(1)^3 = 4\pi$$

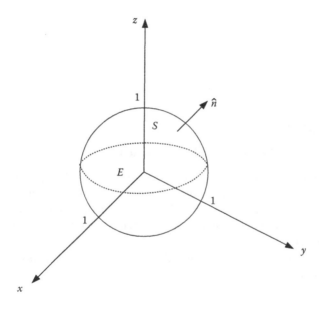

Figure 11.68 Flux out of a sphere.

The right-hand side of Equation 11.38 gives

$$\iint_S (\bar{F}.\hat{n})\,dS = \iint_S \langle x, y, z \rangle . \frac{\langle x, y, z \rangle}{\sqrt{x^2 + y^2 + z^2}}\,dS$$

$$\iint_S \frac{x^2 + y^2 + z^2}{\sqrt{x^2 + y^2 + z^2}}\,dS$$

Now to parameterize the surface of the sphere with $\rho = 1$, let

$$x = \sin\varphi\cos\theta$$
$$y = \sin\varphi\sin\theta$$
$$z = \cos\varphi$$

$0 \leq \theta \leq 2\pi$ and $0 \leq \varphi \leq \pi$

$x^2 + y^2 + z^2 = \rho^2$ and $\rho = 1$

This gives the following:

$$= \iint_S 1 \left| \frac{\partial(x,y,z)}{\partial(\varphi,\theta)} \right| dS$$

$$= \int_0^{2\pi} \int_0^{\pi} (\sin\varphi) \, d\varphi \, d\theta$$

$$= \int_0^{2\pi} \left[-\cos\varphi \right]_0^{\pi} d\theta$$

$$= \int_0^{2\pi} 2 \, d\theta = 2 \times 2\pi = 4\pi$$

So the left-hand side equals the right-hand side as required.

11.10   Applications

Example 11.33: Fluid Dynamics of Smoke Flow

In fluid dynamics one of the main tasks is to find the velocity field describing the flow in a given region. One can make use of the laws of mechanics, that is, the conservation of mass to derive the continuity equation.

Any "small" fluid element can be assigned a velocity $v(x,t)$ and average density $\rho(x,t)$.

Considering a volume V bounded by a surface S, the mass inside the volume is given by

$$\int_V \rho \, dV$$

The rate of decrease of mass in the volume is

$$V = -\frac{d}{dt} \int_V \rho \, dV = -\int_V \frac{\partial\rho}{\partial t} \, dV \tag{11.39}$$

This must be equal to the total rate of mass flux out of V if mass is conserved. The rate of mass flux out of any small element dS of S is given as $\rho v.dS$. Integrating over the whole surface gives the rate of mass flux out of V as

$$= \int_S \rho v.dS$$

Now using the divergence theorem this can be written as a volume integral as

$$\int\limits_{V} \nabla.(\rho v)\,dV \qquad\qquad (11.40)$$

This then implies that for mass to be conserved, Equations 11.39 and 11.40 must be equal for any volume V giving the continuity equation as

$$\frac{\partial \rho}{\partial t} + \nabla.(\rho v) = 0$$

For fluids that are incompressible, ρ is a constant and so this reduces to the following:

$$\nabla.(\rho v) = 0$$

In any flow region, the flow equations are solved to a set of conditions that act at the boundary.

Applications in smoke control situations occur when the smoke temperature of the smoke reservoir does not change. In order to maintain the smoke layer height, the amount of smoke extracted from the smoke reservoir equals the amount of smoke flow into the reservoir from the fire plume, that is, $\nabla.(\rho v) = 0$.

Example 11.34: Application in Thermodynamics

Heat engines such as automobile engines operate in a cyclic manner, adding energy in the form of heat in one part of the cycle and using that energy to do useful work in another part of the cycle. Since work is done only when the volume of the gas changes, the PV diagram gives a visual interpretation of work done. For a cyclic heat engine process, the PV diagram will be a closed loop. The area inside the loop is a representation of the amount of work done during a cycle.

Find the work done by the engine cycle C, where P is the pressure and V is the volume as shown in Figure 11.69.

Parameterization of the cyclic path gives

$$V = 5\cos(-t) + 50$$
$$P = 5\sin(-t) + 30$$
$$0 \le t \le 2\pi$$

The formula for the work done is given by using Equation (11.8),

$$\text{Work} = \int\limits_{C} P.dV$$

$$= \int\limits_{0}^{2\pi} (-5\sin(t) + 30)(-5\sin(t))\,dt$$

$$= \int\limits_{0}^{2\pi} (25\sin^2(t) - 150\sin(t))\,dt = 25\pi\,\text{J}$$

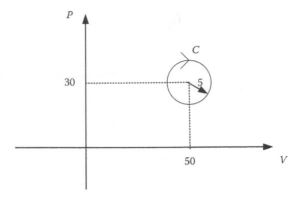

Figure 11.69 The closed path for the engine cycle.

Example 11.35: Velocity of the Earth's Wind

The vorticity plays an important role in fluid dynamics. For the earth's wind, it is defined as the curl of the wind velocity (i.e., curl $\bar{v}$):

$$\text{curl } \bar{v} = \nabla \times \bar{v} = \begin{vmatrix} \underline{i} & \underline{j} & \underline{k} \\ \dfrac{\partial}{\partial x} & \dfrac{\partial}{\partial y} & \dfrac{\partial}{\partial z} \\ u & v & w \end{vmatrix} = \underline{i}\left(\dfrac{\partial w}{\partial y} - \dfrac{\partial v}{\partial z}\right) + \underline{j}\left(\dfrac{\partial u}{\partial z} - \dfrac{\partial w}{\partial x}\right) + \underline{k}\left(\dfrac{\partial v}{\partial x} - \dfrac{\partial u}{\partial y}\right)$$

Since, in general, the horizontal velocities are much larger than the vertical velocities and vertical scales are much smaller than the horizontal scales, in the x and y components of the above expression we can neglect the terms in the vertical velocity, giving

$$\text{curl } \bar{v} \sim \underline{i}\left(-\dfrac{\partial v}{\partial z}\right) + \underline{j}\left(\dfrac{\partial u}{\partial z}\right) + \underline{k}\left(\dfrac{\partial v}{\partial x} - \dfrac{\partial u}{\partial y}\right)$$

The first two terms $\dfrac{\partial v}{\partial z}$ and $\dfrac{\partial u}{\partial z}$ have typical magnitudes $\dfrac{(10ms^{-1})}{(10^{4}m)} \sim 10^{-3}s^{-1}$.

The third term $\dfrac{\partial v}{\partial x} - \dfrac{\partial u}{\partial y}$ has a typical magnitude $\dfrac{(10ms^{-1})}{(10^{6}m)} \sim 10^{-5}s^{-1}$.

But for irrotational flows, the vorticity is 0 (i.e., $\nabla \times \bar{v} = 0$) and then the velocity can be represented as the gradient of a scalar function. This is known as the velocity potential f such that $\bar{v} = \nabla f$.

Example 11.36: Use of Green's Theorem to Calculate Areas

Green's theorem can be used to compute areas of regions as shown in Figure 11.70.
Green's theorem (Equation 11.13) is given by

$$\oint_C M(x,y)\,dx + N(x,y)\,dy = \iint_D (N_x - M_y)\,dA$$

The closed integral is given as

$$\oint_C x\,dy = \iint_D (N_x - M_y)\,dA = \iint_D (1-0)\,dA = \iint_D dA = \text{Area}\,(D)$$

or it can be given as

$$-\oint_C y\,dx = \iint_D (N_x - M_y)\,dA = \iint_D (0--1)\,dA = \iint_D dA = \text{Area}\,(D)$$

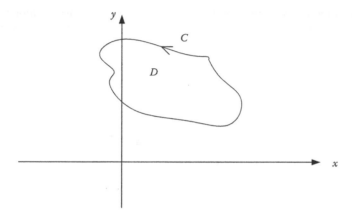

Figure 11.70 Region D bounded by a curve C.

Also, it can be shown that a combination of the above two, that is,

$$\frac{1}{2}\oint_C x\,dy - y\,dx = \iint_D \left(\frac{1}{2} - -\frac{1}{2}\right)\,dA = \iint_D dA = \text{Area}\,(D)$$

This principle has been used to find areas of shapes as one draws a curve around the region. It uses $\oint_C x\,dy$ and keeps a track of this quantity and adds it up to calculate the area.

Some software applications have been developed that allows the calculations of areas of regions on maps using this type of principle.

Problems

11.1 If $\bar{r}(t) = 3t\underline{i} + t^2\underline{j} - 2t^3\underline{k}$, calculate $\bar{r}'(t)$ and $\bar{r}''(t)$.

11.2 Given that $\bar{r}'(t) = \underline{i} + 3t\underline{j} - 6t^2\underline{k}$ and $\bar{r}(0) = \underline{i} + \underline{j} + \underline{k}$, find $\bar{r}(t)$.

11.3 Suppose the density of a wire is given by $d(x,y) = x^2$. Compute its mass when the wire is the semicircle of radius 1 as shown in Figure 11.71.

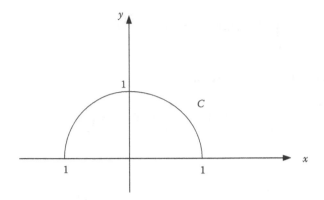

Figure 11.71 A wire in the form of a semicircle.

11.4 A force field is given by $\bar{F}(x,y) = \langle x, y^2 \rangle$, a particle travels along a curve C given by $y = x^2$ as shown in Figure 11.72. Calculate the work done by the particle.

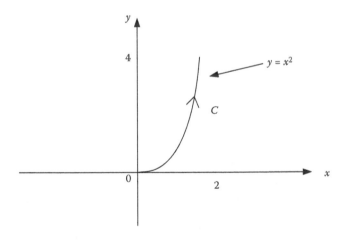

Figure 11.72 Path traveled by the particle.

11.5 Verify Stokes' theorem for the vector field $\bar{F} = \langle 4z, -2x, 2x \rangle$, where C is the intersection of $x^2 + y^2 = 1$ and $z = y + 1$.

11.6 Suppose $\bar{F} = \langle x, y, 0 \rangle$ with E is the cylinder of radius 4 cm and height 5 cm as shown in Figure 11.73. For the vector field verify that

$$\iiint_E \operatorname{div}\bar{F}\, dV = \iint_S (\bar{F}.\hat{n})\, dS = 160\pi$$

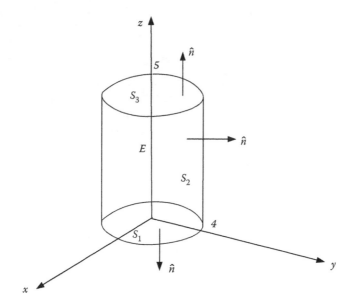

Figure 11.73 Flux coming out of a cylinder.

Answers

Chapter 1

1.1 a. $A = \dfrac{3.6V}{L}$ b. $H = \dfrac{F-G}{7}$ c. $k = \alpha\rho c$ d. $t = \dfrac{h-v^2}{g}$

e. $d = 2\sqrt{\alpha t}$ f. $T = \left(\dfrac{I}{\varepsilon\sigma A}\right)^{\frac{1}{4}}$ g. $t = \dfrac{s\tau}{H^2}\sqrt{\dfrac{H}{g}}$ h. $v = \sqrt{\dfrac{2E}{m}}$

i. $h = \dfrac{3+5g}{g-1}$ j. $\dot{Q} = \left(\dfrac{U}{0.96}\right)^3 H$

1.2 a. $\dot{Q} = r\left[\dfrac{H(T_j - T_0)}{5.38}\right]^{\frac{3}{2}}$ b. 26649.5 kW

1.3 a. $a = 3$ $b = -1$ b. $x = 1.5$ $y = 0.5$ c. $x = 0.54$ or -5.54
d. $x = 3.16$ or -1.16

1.4 a. $t = 10.6$ s, $\theta = 81.5°$ b. $V = 9.65 \text{ms}^{-1}$

1.5 $\bar{x} = 11.04$ $\sigma = 3.61$

Chapter 2

2.1 a. $\dfrac{1}{4}$ b. $\dfrac{2}{13}$ c. $\dfrac{7}{13}$

2.2 $\dfrac{5}{14}$

2.3 a. 0.63 b. 0.97

2.4 $P(0) = \dfrac{1}{8}$, $P(1) = \dfrac{3}{8}$, $P(2) = \dfrac{3}{8}$, $P(3) = \dfrac{1}{8}$

2.5 $P(B\backslash D) = 0.62$

2.6 $E(X) = \dfrac{13}{6}, \sigma = 0.68$

2.7 $E(t) = 2.7$ years, $\sigma = 0.84$ years

2.8 $\dfrac{1}{407}$

Chapter 3

3.1 $\hat{u} = \dfrac{1}{\sqrt{21}} \langle 2, -1, 4 \rangle$

3.2 $\theta = 129.05°$

3.3 $\alpha = -2$

3.5 $-3\underline{i} - 5\underline{j} - 7\underline{k}$

3.6 $4\underline{i} + 9\underline{j} + 5\underline{k}$

3.7 $-3x + 5y + 2z = -3$

3.8 a. 1

3.9 $V_R = 0.57ms^{-1}$ $\theta = 48.1°$

Chapter 4

4.1 a. -13 b. -154

4.2 $k = \dfrac{6}{5}$

4.3 a. $\begin{pmatrix} 6 & 3 & 6 \\ 4 & 2 & 5 \end{pmatrix}$ b. $\begin{pmatrix} -2 & -1 & 4 \\ -2 & 4 & -9 \end{pmatrix}$

4.4 a. $\begin{pmatrix} 9 & -8 \\ 8 & -1 \\ 10 & -15 \end{pmatrix}$ b. $\begin{pmatrix} -16 & -75 \\ 25 & 34 \end{pmatrix}$ c. $\begin{pmatrix} 17 & 14 & 20 \\ 14 & 13 & 15 \\ 20 & 15 & 25 \end{pmatrix}$

4.5 a. $x = 2\ y = 3\ z = -1$ b. $x = 3\ y = 1\ z = 2$

4.6 $i_1 = 2 \quad i_2 = 2 \quad i_3 = 1$

4.7 $\lambda_1 = 1 \quad e_1 = \begin{pmatrix} -1 \\ 1 \\ 0 \end{pmatrix}, \quad \lambda_2 = 2 \quad e_2 = \begin{pmatrix} -3 \\ -3 \\ 1 \end{pmatrix}, \quad \lambda_3 = 21 \quad e_1 = \begin{pmatrix} 1 \\ 1 \\ 0 \end{pmatrix}$

4.8 a. $\dfrac{1}{6}\begin{pmatrix} 1 & 1 \\ -3 & 3 \end{pmatrix}\begin{pmatrix} 1 & 0 \\ 0 & 5 \end{pmatrix}\begin{pmatrix} 3 & -1 \\ 3 & 1 \end{pmatrix}$ b. $A^{10} = \begin{pmatrix} 4882813 & 1627604 \\ 14648436 & 4882813 \end{pmatrix}$

4.9 S = 11 minutes, F = 19.9 minutes, B = 21 minutes

Chapter 5

5.1 a. $-2 \pm j$ b. $\dfrac{1}{4} \pm \dfrac{\sqrt{55}}{4}j$ c. $-1 \pm \dfrac{\sqrt{33}}{3}j$

5.2 a. $4 + 7j$ b. $-j$ c. $-7 + 11j$ d. $\dfrac{13}{5} - \dfrac{1}{5}j$

5.3 $a = 9,\ b = \dfrac{1}{3};\ \ a = \dfrac{1}{4},\ b = 12$

5.4 a. $1.81 + 0.42j,\ -1.27 + 1.36j,\ -0.55 - 1.78j$

  b. $1.21 - 0.17j,\ 0.54 + 1.09j,\ -0.87 + 0.85j,\ -1.08 - 0.57j,\ 0.21 - 1.20j$

  c. $5.80 - 1.65j,\ -5.80 + 1.55j$ d. $2.60 - 1.5j,\ 3j,\ -2.60 - 1.5j$

5.5 a. $z = 32 + 46.91j$ b. $i = 0.119 + 0.175j,\ |i| = 212mA$

5.6 $a_1^2 < 4a_2$

Chapter 6

6.1 a. $12x^2 - 10x + 7$ b. $12\sqrt{x} + 40$ c. $xe^{3x}(3x + 2)$

  d. $\dfrac{x\cos x - \sin x}{x^2}$ e. $15x^2(x^3 + 1)^4$

6.2 (2,0) and (–2,0) both are maximum points

6.3 b. r = 11.8 cm, minimum

6.4 a. $x^4 + 3x^3 - 5x^2 + 4x + C$ b. $\dfrac{6}{5}x^{\frac{5}{2}} + \dfrac{10}{3}x^{\frac{3}{2}} + C$

  c. $\dfrac{x^4}{4} - \dfrac{1}{x} + C$ d. $2(x^2 + 5)^{\frac{3}{2}} + C$ e.   1.106

  f. $\dfrac{xe^{4x}}{4} - \dfrac{e^{4x}}{16} + C$ g. $\dfrac{1}{2}e^x(\sin x - \cos x)$

  h. $\dfrac{x^{n+1}}{n+1}\ln x - \dfrac{x^{n+1}}{(n+1)^2} + C$

6.5 100 J

6.6 3.125 years

Chapter 7

7.1 $y^2 = x^2 + 1$

7.2 $\ln(y-3) = \dfrac{x^2}{2} + 5x + C$

7.3 $\dfrac{y^2}{2} + y = \ln x + C$

7.4 $y = \dfrac{x^3}{2} + Cx$

7.5 $y = \dfrac{x^5}{5(1-x^2)} + \dfrac{1}{1-x^2}$

7.6 $y = \alpha e^{-x} + \beta e^{2x} - \dfrac{x}{2} - \dfrac{3}{4}$

7.7 $y = \alpha e^{5x} + \beta x e^{5x} + \dfrac{10}{25}$

7.8 $y = e^{-2x}(-0.5 \cos x - \sin x) + 0.5 e^{3x}$

7.9 $T = \dfrac{1}{5}(6e^{5t} - 1)$

Chapter 8

8.1 $\dfrac{2}{t^3}$

8.2 a. $y = -\dfrac{1}{9}e^{-2t} + \dfrac{10}{9}e^{7t}$ b. $T = \dfrac{1}{2}(1 + e^{-6t})$

c. $y = 1 + e^t$ d. $x = \dfrac{1}{54}(109 + 18t - e^{18t})$

8.3 a. $i = \dfrac{V_0}{R}\left(1 - e^{-\frac{R}{L}t}\right)$ b. $i = \dfrac{V_0 wL e^{-\frac{R}{L}t}}{R^2 + w^2 L^2} + \dfrac{V_0}{R^2 + w^2 L^2}(R \sin wt - wL \cos wt)$

8.4 $x = \dfrac{C}{2k}(e^{kt} - e^{-kt})$ $y = \dfrac{C}{2k}(e^{kt} + e^{-kt}) - \dfrac{C}{k}$

Chapter 9

9.1 a. $a_0 = 1$, $a_n = \begin{cases} 0 & n = even \\ -\dfrac{4}{n^2\pi^2} & n = odd \end{cases}$, $b_n = \begin{cases} -\dfrac{2}{n\pi} & n = even \\ \dfrac{2}{n\pi} & n = odd \end{cases}$

b. $a_0 = 1$, $a_n = 0$ all n, $b_n = \begin{cases} 0 & n = even \\ \dfrac{2}{n\pi} & n = odd \end{cases}$

9.2 a. $c_n = \dfrac{1}{jn\pi}\left[3 - 5(-1)^n\right]$ b. $a_0 = 0$, $a_n = 0$, $b_n = \begin{cases} -\dfrac{4}{n\pi} & n = even \\ \dfrac{16}{n\pi} & n = odd \end{cases}$

Chapter 10

10.1 $\dfrac{\partial f}{\partial x} = 3$, $\dfrac{\partial f}{\partial y} = 13$

10.2 a. $xy \cos xy + \sin xy$

b. $x^2 \cos xy$ c. $f_{xy} = -x^2 y \sin xy + 2x \cos xy$

d. $f_{yx} = -x^2 y \sin xy + 2x \cos xy$

10.3 $-\dfrac{15}{\sqrt{10}}$

10.4 -18

10.5 $\dfrac{16}{3}$

Chapter 11

11.1 $\bar{r}'(t) = \langle 3,\ 2t,\ -6t^2 \rangle$, $\bar{r}''(t) = \langle 0,\ 2,\ -12t \rangle$

11.2 $\bar{r}(t) = (1+t)\underline{i} + (1+1.5t^2)\underline{j} + (1-2t^3)\underline{k}$

11.3 $\dfrac{\pi}{2}$

11.4 10

Index

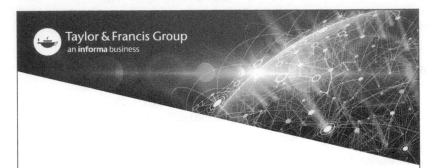

Taylor & Francis Group
an **informa** business

Taylor & Francis eBooks

www.taylorfrancis.com

A single destination for eBooks from Taylor & Francis with increased functionality and an improved user experience to meet the needs of our customers.

90,000+ eBooks of award-winning academic content in Humanities, Social Science, Science, Technology, Engineering, and Medical written by a global network of editors and authors.

TAYLOR & FRANCIS EBOOKS OFFERS:

A streamlined experience for our library customers

A single point of discovery for all of our eBook content

Improved search and discovery of content at both book and chapter level

REQUEST A FREE TRIAL
support@taylorfrancis.com

Routledge
Taylor & Francis Group

CRC Press
Taylor & Francis Group

und by CPI Group (UK) Ltd, Croydon, CR0 4YY

24/10/2024

01778291-0008